The Scientific Theory of God

A Bridge Between Faith and Physics

By

Shelton W. Riggs, Jr.

The Scientific Theory of God

A Bridge Between Faith and Physics

Copyright © 2007 by Shelton W. Riggs, Jr

ISBN-13: 978-1-44862-004-3

ISBN-10: 1-44862-004-X

This book is dedicated to

my late, beloved wife, Robin,

whose original ideas,

suggestions and analysis,

made this work possible.

About The Author

Shelton W. Riggs, Jr. earned undergraduate (University of Texas) and graduate (Vanderbilt) degrees in both Physics and Mathematics.

Professionally, he has consulted as both a hardware and software design engineer to numerous Fortune 500 companies for a wide range of scientific applications. He helped solve several scientific problems for US Army, Air Force and Navy.

Other interests include theoretical physics including quantum mechanics, relativistic mechanics and theoretical mathematics (especially the mystery of prime numbers).

Hobbies include dancing, karaoke, juggling, playing keyboards, writing songs, and writing poetry.

Other Works By Author

Nature of the First Cause – The Discovery of What Triggered the Big Bang contains the formal scientific theory of how the universe got started. It lays down the mathematical foundation for the creation theory put forth in this book, "The scientific Theory of God". It resolves the asymmetry problem of physics. It solves the two main cosmological problems by identifying both dark energy and dark matter. This theory predicts the correct order of magnitude for the number of galaxies and stars in the universe revealed by the Hubble ultra deep field results. It uncovers two entangled parallel worlds consisting of negative antimatter and positive matter. It explains the accelerated expansion of both matter and negative antimatter. It predicts the distance between matter and negative antimatter to be the Schwarzschild diameter of the expanding universe.

An Alternate Lorentz Invariant Relativistic Wave Equation offers an invariant form which differs from both the Dirac equation as well as the

Klein-Gordon equation. Unlike, Schrodinger's non-relativistic wave equation, both the Dirac as well as the Klein-Gordon equation predict wave functions which do not collapse when applied to free systems at rest. On the other hand, Schrodinger's equation predicts wave functions that do collapse when applied to free systems at rest. The alternate relativistic wave equation offered by the author follows Schrodinger's philosophy that if a free system is at rest, then it is a particle with a collapsed wave function.

The Origin of the Planck Mass, Planck Length and Planck Time presents solutions of a system composed of two identical photons which are trapped in each other's gravitational field. The solution applies to any pair of identical particles having zero rest mass. Two solutions were derived. One solution was found by treating two photons as point particles. The quantum mechanical solution came about by treating the two photons as waves. In both solutions, the predicted distance between photons was found to be proportional to the Planck length. The period of the photon's orbit was proportional to the Planck time and the mass energy of each photon is proportional to the Planck mass. The concept of a Planck length, Planck mass and Planck time all emerge from this single model.

Primal Proofs offer several proofs that deal with prime numbers. A proof by contradiction of Goldbach's binary conjecture that every even natural number greater than two (2) can be expressed as the sum of two (2) primes is given. A proof of Goldbach's ternary conjecture that all natural numbers greater than five (5) are the sum of three (3) primes via the binary proof is presented. A proof by construction (utilizing the proof of the binary conjecture) of the twin prime conjecture is offered. A proof of the Riemann hypothesis by deduction is presented. A proof that any prime greater than three (3) is the mean of two other primes is presented. A proof is offered that any even number greater than twelve (12) satisfies Goldbach's binary conjecture in a plurality of ways. Two entangled formulas that generate all the primes beyond the second prime ($P_2 = 3$) are developed and summarized.

Acknowledgments

I acknowledge God for providing me with all my teachers who are all partially responsible for this work. Special thanks goes out to all my science and mathematics teachers.

I acknowledge my parents for providing a secure and nurturing environment that initially made my learning fun.

I acknowledge my country for providing me with all my freedoms, especially the freedom of speech.

I acknowledge every author in the reference section of this book for providing both ideas and data.

Preface

This manuscript is intended to provide the reader with basic scientific understanding, interpretation, clarification and answers about concepts and beliefs associated with a Supreme Being. The ideas that are developed are based on current scientific theory and the standard model of physics. The study and/or belief in God may now rest upon a firm scientific foundation. This new basis has revealed new insights and uncovered surprising relationships between the scientific definitions of both God and life. Moreover, this book offers a scientific creation theory and shows how it is compatible with both the big bang as well as evolutionary theory.

The body of definitions lend themselves to the application of the scientific method. The analysis and results are based on how these definitions are related and verified by observation and experiment. Conclusions presented in this manuscript are backed by scientific proofs which will be disclosed or clarified upon written request.

An analytical definition of life is presented. A model for the behavior of living matter (bioenergy) has been developed. The model has been extended to include the behavior of human bioenergy in terms of perception, decision and action. These concepts combined with the operation of short and long-term memory explain both human consciousness and how the mind controls the body. This model also includes how any desired behavior (provided it does not go against survival) may be achieved.

This information has also produced a special set of terminology which has required new and precise definitions. A glossary containing these definitions has been provided. The reader is advised to be aware of these special terms before reading the text and should make use of this glossary if the text should require clarification.

The fundamental physical laws, basic units, physical constants and basic elementary particles have been included immediately after the glossary.

Definitions in the text are presented in bold type that are not section or chapter titles. Glossary definitions are in bold caps. Section titles are in bold type.

In the "Fundamental Physical Laws" section, vectors in equations are in bold type.

Table of Contents

Chapter 1
Scientific Preliminaries

Let us begin by examining the value of experimental science. It is the only tool which allows human beings to reliably predict future events. It also enables us to accurately describe the past and present. This ultimately translates into knowledge. It gets its answers directly from nature herself in the form of unbiased data. It is precisely this "knowledge about the behavior of nature" that will ultimately decide the fate of the earth's life forms. As everyone is aware, the application of knowledge is a double edged sword. Today's fund of knowledge has not only produced the highest living standards on earth, it has also produced the most destructive forces on earth. Ironically, while high standards exist for some human life, other species succumb to extinction.

Is there some knowledge which could guarantee long term preservation and survival of all earth's life forms? What are the weaknesses in today's fund of essential survival knowledge?

Can the scientific method be applied to fundamental questions of theology such as:

(1) What is God?

(2) Does God exist?

(3) Is God alive?

(4) Did God create the universe?

(5) Did God create Man?

(6) What is a soul? What is a spirit?

(7) Are there any limits on what can be known about God?

What can science now say about fundamental philosophical questions such as:

(1) What is life? What distinguishes living from non-living?

(2) How does one distinguish between truth and falsity?

(3) Is there a purpose for life?

(4) Is there evidence for life after death?

(5) How can reality be defined?

(6) Is there a valid model for consciousness and how the human mind works?

(7) Is there a logically consistent basis for human morality?

The second fundamental question of theology "Does God exist?" merits a bit further discussion. One must realize that the question begs for a definition. In other words, if men of God could tell scientists what God is, then science could possibly determine whether or not God exists. Most classic definitions of God contains several persistent themes such as:

(1) God is the first cause or creator of the universe.

(2) God is all powerful.

(3) God is present everywhere.

(4) God is all knowing.

(5) God is the designer of the universe.

(6) God is the controller of the universe.

(7) God is the loving giver of life.

Creation Paradox

If God created the Universe, then who created God? The Creation Paradox is: If nothing created God, then God (uncreated) cannot exist. However, if something created God, then something exists which is greater than God.

Evolution Paradox

Did God create man or did he evolve by chance? The evolution paradox is: If God created man, then how can biology claim that mankind evolved. If biological fossil records imply that man evolved, then how could God have created man?

With a few simple definitions, the scientific method can be applied to all of these questions and paradoxes with some surprising results.

We will now turn to a brief and simple summary of modern scientific beliefs and the methods of physical science.

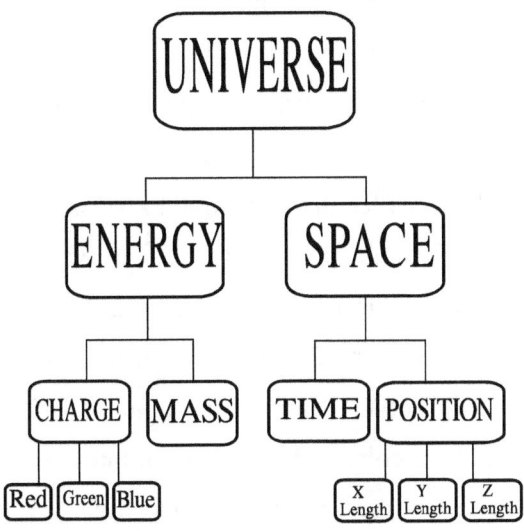

Figure 1   Conceptual Universe

The Universe

When science speaks of the universe, it means the entire observable universe. If you say, but what about other as yet unheard of or undiscovered universes? The entire universe conceptually includes all those (including multiverse(s)) which can be measured, in its definition. This means we may speak about the universe as a unit since all others may conceptually be summed up into one. Figure 1 is a depiction of how the universe can be described and predicted using only four measurable quantities which are electric charge, mass, time and position (length). Three dimensional position include the three lengths of x, y and z. Microscopic color includes red, green and blue carried by the three fractional electric charged quarks of both a proton and neutron. Positive electrical charge of atomic nuclei results from equal amounts of red, green and blue color and is always an integer times the charge on the proton. The electron has the smallest mass which can carry a single negative electrical charge.

The experimentally observed universe may be thought of as consisting of two, but yet inseparable components. The first component is energy. The second component is space. The view of the universe may then be energy in space.

The established big bang theory of the universe (which much experimental evidence confirms)

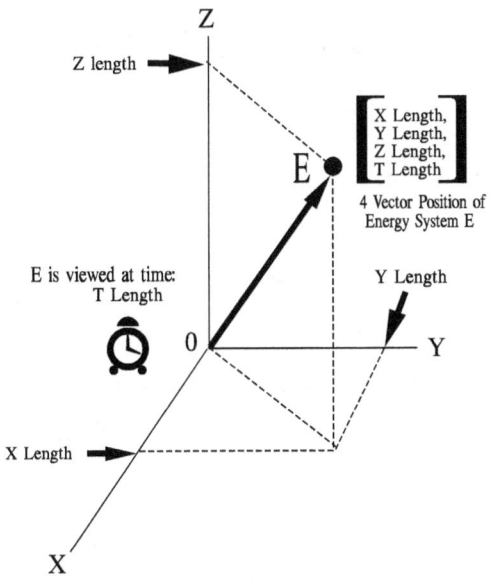

Figure 2 Four Dimensional Space

shows that at some past time, all the energy of the universe was concentrated into a small volume and then proceeded to unfold into what is observed to be our present day expanding universe.

Figure 2 shows the four vector position (x,y,z,t) of an energy system and includes where and when it was. It turns out that the relativistic four vector position of a system contains the time of when it was.

Fundamental Physical Descriptors

The energy part of the universe may be described by only two quantities. These are the concept of mass and the concept of charge. Mass is equivalent to stored energy and has inertia and takes up space. Electric charge comes in two flavors, the plus type and the minus type. Like charges repel, unlike charges attract. Charge always resides upon some mass which fundamentally (protons and neutrons) consists of quark trios. Each quark can contain a red, green or blue color which keep the quark trios of both protons and neutrons together. Moreover, a quark can contain a $-1/3$ or a $+2/3$ electric charge. The electron and positron have the smallest mass that can carry a -1 and $+1$ electron charge respectively. Integral charges (0, 1, 2, ...

etc.) always have no color which is the same as having equal amounts of red, green and blue color. Mass is measured with a balance scale. Charge is measured with a voltmeter.

Similarly, the space part of the universe may be described by only two quantities. These are the concept of position (x length, y length & z length) and the notion of a length of time t (t length). Thus, the position of an energy system can be represented by a four dimensional (x,y,z,t) position with its four components of length. Spatial length is measured with a ruler. Temporal length (time) is measured with a clock. In a nutshell, only four fundamental physical quantities which are defined in terms of measurement and of which all other defined and measureable quantities of science are derived. Again, energy descriptors are mass and charge, while the space descriptors are position (lengths) and time.

Fields of Energy

There are four fundamental fields of force. Fields in space are defined in terms of the energy descriptors of mass or charge existing in space described by both position and time.

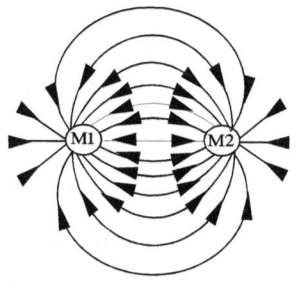

Gravitational Field of Two Masses: M1 & M2

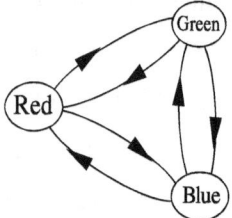

Nuclear Force Field of Three Colors: Red, Green & Blue

Figure 3   Gravitational & Nuclear Fields

Figure 3 shows the gravitational field of two attracting masses and the strong nuclear force field between three attracting color quarks.

The presence of mass originates the gravitational force field. The presence of colored quarks originates the strong nuclear force field. The presence of charge originates the electric force field. The motion of charge originates the magnetic force field.

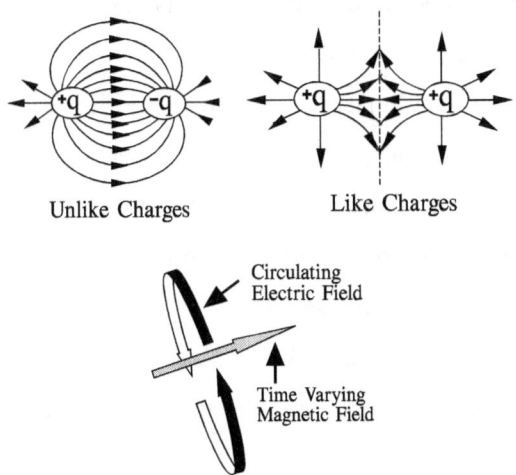

Figure 4 Electric Fields

Electric and magnetic fields may also be produced by induction. If a magnetic field is changing with respect to time, it induces a corresponding electric field. Moreover, induced magnetic fields are produced by time varying electric fields.

Figure 4 shows the electric field of force between unlike charges, like charges, as well as induced electric fields.

The electric field serves to hold unlike charges together, repel like charges, and is responsible for all chemical behavior.

The gravitational field is responsible for firing up the sun and stars, holding us to the earth, holding moons to the planets, holding planets to the stars, holding stars to the galaxy and finally holding galaxies to the universe.

Many of you that have played with magnets have directly experienced the magic of a magnetic field.

Figure 5 depicts magnetic fields which are caused by charge in motion and an induced magnetic field caused by a time varying electric field. A charge moving away from the reader causes the circulating magnetic fields. In a bar magnet, electrons (which carry a negative charge) are all aligned and the magnetic field of each contributes to the entire magnetic field of the magnet.

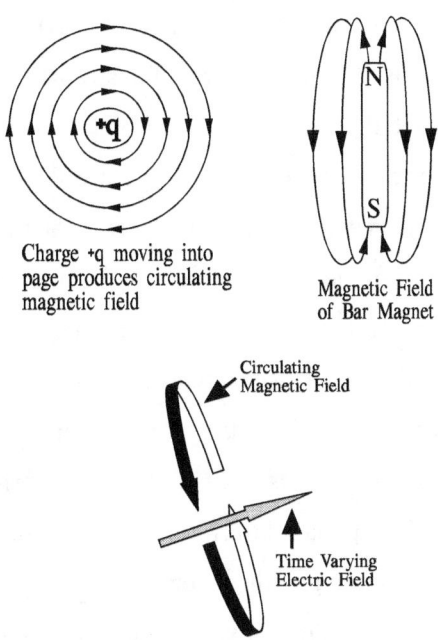

Charge +q moving into
page produces circulating
magnetic field

Magnetic Field
of Bar Magnet

Circulating
Magnetic Field

Time Varying
Electric Field

Figure 5 Magnetic Fields

The strong nuclear force field is the strongest field and is responsible for holding (by gluons) the quark trios together which compose both neutrons and protons. Nuclei of atoms are composed of protons and neutrons held (by pions) together as atomic nuclei. If it were not for the nuclear force, nuclei possessing two or more protons (protons have a plus charge) would repel each other and immediately fly apart (like charges repel). Thus, if it weren't for the nuclear force, the atomic elements could not exist.

Thus, the force fields of energy are the electric, magnetic, gravitational and strong nuclear fields. In the standard model literature, one will encounter the electroweak force responsible for radioactivity. However, the electroweak field may be considered a combination of both strong nuclear and electromagnetic fields. All fields of energy are described by spatial descriptors (length and time) and are associated with a corresponding potential energy. Thus, there exists magnetic potential energy, electric potential energy, gravitational potential energy and nuclear potential energy.

All these force fields are continually influencing the behavior of the universe. Additionally, every field of force contains a specific amount of field potential energy and is spread over space and time. Moreover, the strength of any kind (gravitational, electric, magnetic or strong nuclear) of force field at any point in space is the sum of like kind fields caused by all sources in the universe at that point. This means that the strength of each force field can be represented by a single number at every point in space.

It must be mentioned that in the standard model, matter particles are known as fermions which have half integral values of Planck's constant divided by 2π as their axial spin, which is the smallest possible value of angular momentum. On the other hand, there exists field particles called bosons which have integral values of Planck's constant divided by 2π as their axial spin and are responsible for forces and potential energy between the matter particles. Thus, the fields of force also belong to the energy part of the universe and are all defined in terms of the four fundamental descriptors.

Subsystems of the universe will always contain specific amounts of energy so that when we speak of subsystems like the moon or the bird or the human being, we ultimately are referring to energy representations of the moon, or the energy representation of the bird or the energy representation of the human being. The motions of these subsystems in space give rise to kinetic energy when subjected to the potential energy of the fields of force.

The Fundamental Law of Nature

Energy cannot be created nor destroyed. In all the experiments to date, the amount of total energy (kinetic plus potential) of any subsystem remains constant. In any experiment, one form of energy transforms itself into another form. The left side of Figure 6 depicts the total energy E_T (in terms of material energy E_M and Field energy E_F) before, during and after the big bang. The total material energy, E_M consists of positive matter energy, E_+, negative antimatter energy, E_- and associated gravitational energy, E_G which causes positive matter and negative antimatter to repel each other. The sum of all three energies is precisely zero. Note that E_T (the total energy of the universe) remains constant and is the same as the initial field energy,

E_F.

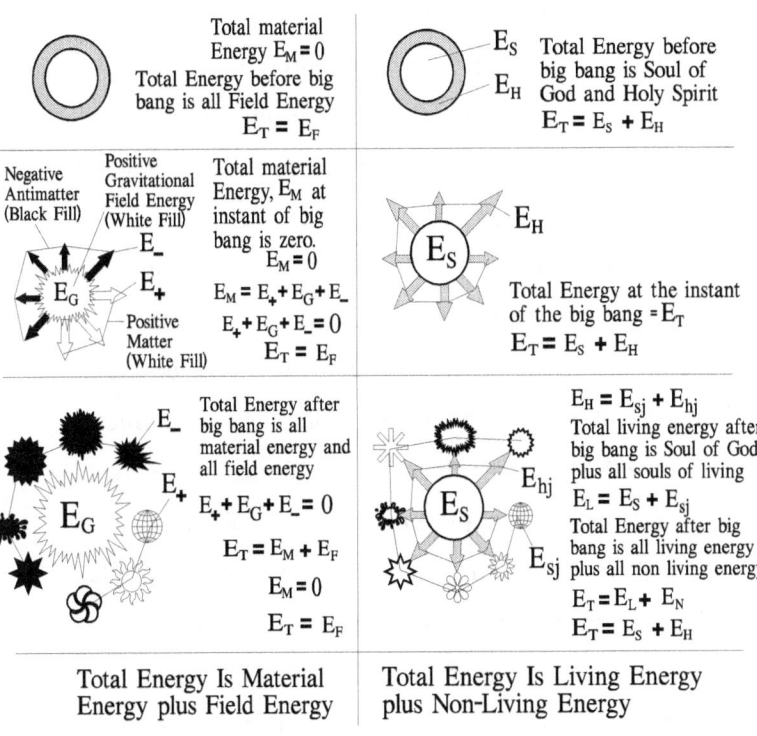

Figure 6   Energy Conservation of Big Bang

Methods of Science

The main reason that science is the only endeavor which is capable of describing and predicting events lies in its method. It relies entirely on careful observation of what nature does under known conditions. All the laws of science as well as all the established scientific theories are founded by experiment. What does this mean? In order to establish scientific laws, the variables which are going to be measured, are defined. The definitions may be arbitrary. The only rules for the definitions are consistency. Once something is defined, it cannot be changed during the experiment or in the description or prediction of the experiment. Repeatable experimentation ultimately reveals relationships between the definitions. The mathematical expression relating the variables which have been defined express the laws.

The collection of laws describing or predicting a broad category of events is a scientific theory. It must be mentioned that even though thousands of experiments may verify a scientific theory, it only takes one experiment to disprove it. Herein lies the strength of the scientific method. The validity of theory is tested via experiment. Either the theory predicts the outcome accurately or it does not. Thus, a scientific theory is a self correcting body of descriptive and predictive knowledge.

Theory and Scientific Theory

There is a big difference between a theory and a scientific theory. A scientific theory is backed by experimental evidence and a set of mathematical equations or statements that relate the definitions of the theory. These equations allow for the theory to predict the outcome of an experiment. A theory which cannot be tested via experiment, is only conjecture. Scientific theory is concerned with the behavior of energy. The ability to describe and predict the behavior of energy is tied up in the equations of a scientific theory. This ability to describe and predict the behavior of energy translates to the ability to survive. Knowledge is Survivability.

Chapter 2
Scientific Concept of God

Classical Concepts of God

The classical characteristics of God include the following concepts. God has been depicted as all powerful, all knowing, and forever existing in time and space. Many religions describe God as the first cause of the universe or the creator of the universe. The underlying theme in all the historical arguments for the existence of God is the concept of the "ultimate or most perfect or the greatest". Thus, even though the classical concepts of God did not offer a scientifically rigorous definition, they certainly pointed the way towards such a definition. All that is needed is a definition which is scientifically sound and which would combine the maximum number of all the classical ideas of a Supreme Being.

Scientific Definition of God

Consider the following definition of God. **God is composed of all energy in the universe**. This definition is very much in accord with all the classical ideas of God. It readily lays down an experimental basis for investigating our ideas of a Supreme Being. The definition unifies both a basic concept of physics and the basic premise of most religious belief. With this scientific definition, the following statements may be made. God is invariable and cannot be created nor destroyed. God occupies all space and time. All things are a part of God. God's original living energy is within all life forms and all life forms are a part of God. Essentially this definition is consistent with all the classical concepts of a Supreme Being.

Universal Spirit and Soul of God

Along with defining the scientific essence of God, in order to be compatible with the classical ideas of a Supreme Being, **God's Universal Spirit is defined to be composed of all Force Field Energy in the Universe**. Even though the definition of God includes all the energy associated with the universal force fields, it is essential to distinguish between total force field energy and total energy. The physical distinction is between the energy which is being influenced (behaving or moving) and the energy which causes the influence (the fields of force). God moves and connects all components of the universe via the Universal Spirit. By definition, God is all field energy plus all material energy.

However, before the big bang, it can be shown that there was no material energy (the same as having equal positive and negative energy components) therefore, the total energy (God) was all field energy before the big bang. The only field energy known which can persist in a non material environment is induced electromagnetic fields. An example is the existence of induced electromagnetic waves of light energy or photons in the vacuum of space.

The only other possibility is that there was no field energy before the big bang which would mean zero total energy before the big bang. This in turn would mean that God could not have existed before the big bang.

Proof By Contradiction That God's Energy Is Non-Zero

Let us assume that before the big bang, there was no field energy. Since there was no material energy, this would mean that the total energy was exactly zero. In other words, this is the same as assuming that God had no material or spiritual substance. It is assumed that anything that is alive must have non-zero energy. Thus, if the total energy was zero, then before the big bang, God was dead.

If the total energy before the big bang was zero, it would always be in balance. There could never be any positive or negative energy fluctuations. Nothing could move since there was no material and nothing could push or pull since there was no field energy. If there was nothing to fluctuate or move, then time could not have a direction or beginning. Thus, a perfect equilibrium would perpetually exist. Nothing could change if nothing existed. There would be no first cause since none of its non-existent effects could ever take place. No energy could ever be transformed since none originally existed. The universe would always be in a perfect symmetrical state of nothingness, at absolute zero temperature, in the lowest possible energy state, with no cause for anything. The basic physical laws clearly predict what happens to energy systems with no forces acting upon them. Systems with no unbalanced forces acting upon them cannot change their energy states. With the assumption that there was no field energy before the big bang, predicts a steady state universe with perpetual nothingness and a non-living God.

All experimental evidence to date confirms that our universe is anything but a steady state system. The repeatable experimental evidence for the big bang has come from several different independent experimental sources. All the data says that it is looking more and more, like the material world and time itself began some fourteen billion years ago, expanding with a continuing accelerated motion, from a very small volume of space. Astronomers and cosmologists all agree that it also has an unknown dark energy (negative energy) component as well as an unknown dark matter (positive energy) component.

Thus, the zero energy God assumption is clearly contradicted. Since there are only two possibilities and we have shown that the zero energy God assumption is incorrect, then God must have a non-zero energy substance. God's substance is equivalent to the field energy before the big bang. The only field energy which can be sustained in a non-material environment (vacuum of space) are the self induced electromagnetic fields. That we are alive, means that our living energy came from these same induced electromagnetic fields which must have existed before creation.

Thus, the following definition is that **The Universal Spirit before the big bang was purely induced electromagnetic field energy with a controlling, living (Mind) component defined to be the Soul of God and a connective (Substance or Body) component defined to be the Holy Spirit**.

Collapse of the Creation Paradox

Let us now examine some of the advantages that the definitions of God and Universal Spirit have revealed. First, the creation paradox collapses. This is because according to the definitions, God cannot be created nor destroyed (conservation of energy). This is also the correct experimental observation. It also extends what can be known about God. For instance, all energy transformations do so via the dictates of the various force fields. Therefore, all energy transformations in the universe occur under the influence of God's universal spirit (total force fields). This is only one simple example of how many of the classic questions may now be considered and answered. Any body of religious beliefs can be made much clearer with the adoption of only a few precise scientific definitions.

Collapse of the Evolution Paradox

That God created life and that life evolved are equivalent statements. The life giving force of the sun is directly achieved through the two competing force fields associated with mass (a fundamental universal descriptor). First, there is the gravitational field associated with the mass of the sun. It not only functions to hold the planets (including our earth) in their respective orbits, but it causes the enormous pressure in the sun's interior necessary for thermonuclear fusion (only possible because of nuclear forces and electromagnetic forces) which in turn are responsible for its life giving radiation. Since God (as the total energy with total force fields) must have caused the big bang and the creation (beginning of motion and time) of the universe including our sun, planets and ourselves, it follows that God transformed part of His living energy into the material world. The laws of life and the laws of evolution are the result of this ongoing spiritual energy transformation from the Holy Spirit in a matter free (before the big bang) space into the living entities with material form (after the big bang). Please refer to the right hand side of figure 6.

Before the big bang, the total energy was only induced electromagnetic field energy (Universal Spirit) and consisted of two interacting parts. These were the connective Holy Spirit (body of God) and the controlling Soul (mind of God) by definition. However, after the big bang, part of the Holy Spirit was transformed into the induced electromagnetic field energy of living souls. The Soul of God remained the same as before creation. Thus, each living soul is a part of the original (before creation and the big bang) Holy Spirit of God. Before any creation, only God was alive. After creation, the development of new souls came from part of the induced electromagnetic field energy of the Holy

Spirit. The rest of the Holy Spirit became the electromagnetic conduit between the Soul of God and each soul of the living. Thus, the big bang is seen to be a spiritual expansion of the Holy Spirit into all living souls as well as the observed accelerated expansion of material energy and field energy which makes up the universe. The right hand side of figure 6 shows how the total energy before the big bang was the energy of God's Soul and the energy of the Holy Spirit. After the big bang, the energy of the Holy Spirit was converted and expanded into all the souls of the living. All living energy is the sum of all living souls plus the Soul of God. Thus, the total energy is seen to be the energy of the living and the energy of the non-living. The Soul of God remains constant both before and after the big bang just as the total energy.

Changes Within the Unchangeable

Energy experimentally observed reveals that all parts of the universal energy system are seen to be in a state of perpetual transformation. Energy subsystems constantly are transformed into other energy forms. There are births and there are deaths. The mass of our sun is steadily transformed into a spectrum of radiative energy. The waters of the oceans are transformed into gaseous clouds. Rain and rivers flow back into the oceans. Cycles of energy conversion rock back and forth in an unending display of dynamic transition. However, amidst all this change, the total energy of the universe remains constant and invariable. Thus, the essence of God (total energy) is invariable.

Chapter 3
Bioenergy

A Question of Life

Before getting into specific descriptions of living systems, let us turn our attention to one of the most fundamental of all philosophical questions. What is the difference between living energy and non-living energy. To put it more simply "What is life?". The question may seem absurd since typically the distinction between what is alive and not alive seems obvious. The researcher finds that what is obvious is that there is no classic basis for determining what is alive and what is not. The philosophical, biological, psychological, medical and even the legal definitions of life are inadequate in the sense of any overall analytically quantifiable, acceptable viewpoint as to what is or what is not alive.

Living and Non-Living

All of us more or less can distinguish between life and non-life. However, when one seeks the specific property or characteristic which determines if something is alive, the distinction falls apart at the atomic level. In other words, a certain atom, say a carbon atom taken from a "living" object is indistinguishable from one taken from a non-living object. This should not be the case especially when it can be demonstrated that both living and non-living objects are constructed of atoms. Yet clearly, even though living matter cannot be distinguished from non-living matter at the atomic level, a basis for the distinction must exist. It turns out that it is a special group of atoms along with special electromagnetic field energy that afford the distinction between the living and the non-living.

The Living Cell

The biologists categorized the two great branches of bioenergy into plants and animals. From the smallest living creature to the largest, a physical characteristic of all bioenergy emerges. Every living structure is made up of units called cells. This is a basic macroscopic component of any form of bioenergy. The cell contains a dynamic microscopic nucleus which is the genetic control center of the cell. Of what then, is this microscopic genetic material composed? Is the stuff of life within this cell nucleus?

Molecular and Electromagnetic Basis of Life

Bioenergy considered on a cell nucleus level reveals an even more striking organizational feature. This is due to the persistence of a certain rather mysterious dynamic molecule which fundamentally makes up the genetic material of all cells. This is the now famous DNA molecule first described by James D.Watson and Francis Crick. This molecule when observed during cell division is seen to be converting other molecules outside the nucleus of the cell to more DNA molecules. This is its fundamental replicating behavior. This is in contrast to other molecules, for example a water molecule contributes nothing to its own structure. In this sense, a water molecule exists, but does not behave. Every living cell on planet earth contains DNA molecules. They always have the shape of a double helix much like a spiral staircase. Moreover, the DNA molecule is always made up of hydrogen, carbon, nitrogen, oxygen, phosphorous and sulfur atoms. This is not to say that all DNA molecules are identical, each has its own characteristics determined from the particular cell taken from a particular organism. Figure 7 illustrates a portion of a DNA molecule.

The DNA variations between different species are much wider than variations from cell to cell of a single multi-celled organism. In fact, every cell of a multi-celled organism contains identical nuclei including its sex cells (containing a unique half nucleus). It is not the individual atom which determines life or non-life. It is the spatially dynamic combination (arrangement and types) of atoms that determines whether it is possible for it to be alive or not alive (bio or non-bio).

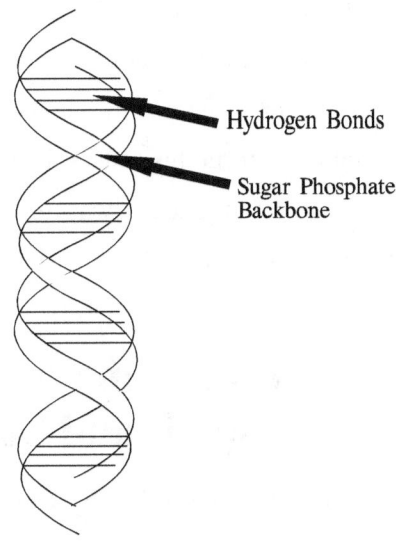

Figure 7   Portion of a DNA molecule

It takes more than the presence of a DNA molecule itself to have life. The molecule must be in a dynamic state such that its structure is undergoing change in time which contributes to the maintenance of its general functional stability.

Obviously, a DNA molecule may be "dead" in the sense that its does not behave or change in time. DNA molecules stay together because of the static electromagnetic fields between electrons and protons of its atoms. DNA molecules behave because of a perturbation by induced dynamic electromagnetic fields onto the static fields holding it together. In turn, this causes the DNA molecule to react or move in such a way to even include self replication behavior.

Electromagnetic Spirit and Soul

What is the mechanism by which atoms interact with each other to form bioenergy? The atoms of molecules are held together via their sharing of electrons. This is only possible because of the static electromagnetic fields which exist between negatively charged electrons and positively charged protons in of atomic nuclei. This force is commonly known as Coulomb force. **Define the body spirit of any living entity as the total electromagnetic fields associated with the molecules that make up and hold the living entity together. Define its soul as the total induced electromagnetic fields that control (by perturbation) the body spirit of any living entity.** It is assumed that the soul receives perceptual electrical signals and sends controlling electrical signals via the body spirit. In other words, the soul controls the body spirit which connects up the material body. It must be kept in mind that even though the qualitative nature of these electromagnetic fields are observed and measured, the quantitative aspects of these fields remain a mystery. In a book entitled "The Fields of Life", Dr. H. S. Burr discusses the significance of measurable controlling electromagnetic "L Fields" present in all life forms. It is these fields that serve as the template for renewing the molecules of cells which make up the protoplasm of all living entities. However, the laws that describe the electromagnetic behavior of even a single living DNA molecule remain unknown.

The Induced Soul

Each atom within a non-living molecule is held in place by electromagnetic fields which results only from the presence and motion of charge. This is only part of the story in the case for living DNA molecules. Recall that electric fields arise from either charge or dynamic magnetic fields. Similarly, magnetic fields may arise from either charge in motion or dynamic electric fields. Ampere's law and Faraday's law (the last two of Maxwell's equations in the electromechanical laws section of this manuscript) describe induced electromagnetic fields even when both charge density and current density are set to zero. It is precisely these dynamic (induced fields) that form the soul (which controls the body spirit) associated with that organism. These induced fields are solutions to the electromagnetic wave equations given at the end of the electromechanical laws section. These induced fields can exist in a matter free (no charge and no current) environment.

Note that this controlling component of bioenergy's electromagnetic field does not arise from the material part of the organism nor of the material part of the universe since it is not caused by charge or charge in motion. When the soul exists as the electromagnetic controller within the material organism, the organism is alive and its structure is preserved. When the soul ceases to control the material organism via the body spirit, the organism dies and its material structure decomposes and the body spirit dissipates. Since the soul is induced, it can transcend the decomposition of the organism's structure as well as the dissipation of the organism's body spirit. The soul is analogous to light (an electromagnetic wave) which propagates through space as a series of induced electric and magnetic fields. The soul cannot have rest mass, as a solution to the electromagnetic wave equations. Thus, souls are capable of moving at the velocity of light in vacuum.

Description of Cells

The induced electromagnetic fields (soul) and electromagnetic fields due to charge and the motion of charge (body spirit) are responsible for the behavior of the cell nucleus. This behavior includes reproduction and metabolism. The nucleus is composed of a set number of chromosomes. It is the structure and number of chromosomes which distinguish one species from another. The nuclei of every complete cell of an organism are identical. The sex cells contain identical half nuclei and each organism has its own representative set of sex cells. The sex cell contains only half the chromosomes that all other non-sex cells have. Normally, the female sex cell is called an egg cell, while the male sex cell is called the sperm cell. The chromosomes are composed of substructures called genes. Furthermore, the gene is a subunit of a DNA molecule. DNA taken from any cell nucleus of an organism is identical and unique to that organism and its identical twin(s) if any.

New Living Souls
Are Inherited

The first cell of any member of any species begins with the unification of male and female sex cells. This unification produces unique DNA molecules formed from the DNA donated by the sex cells of both parents. It must be emphasized, that both male and female sex cells were alive before they were united. The unification of alive, opposite sex cells produces a unique member of any species, that inherited its unique DNA from both parents. The DNA dictates the exact specification of a body with its body spirit. The induced electromagnetic field energy of both sperm and egg are also fused and transformed into a single controlling soul.

A Virus Is Dead and Alive

A virus is not composed of cells, and resembles the nucleus of a cell and contains DNA. When it is outside a cell host, it is dead and its DNA is not dynamic. When it is inside a cell host, it becomes alive and its DNA performs self replication behavior.

Time Cloning

A rather intriguing aspect of bioenergy on a cellular level is a process called cloning. Consider a species containing both female and male sex cells. It is not theoretically possible for the species to survive without females. Surprisingly, it is possible for the species to survive without males.

This may be achieved via cloning. Take a female sex cell and remove the half nucleus turning it temporarily into a blank female sex cell. Take any other cell from that species (either male or female and either from itself or another individual of that species excluding sex cells) and remove the complete nucleus from that cell. Next, implant this complete nucleus into the blank female sex cell and presto, the female sex cell believes itself to be fertilized. Thus, a perfect duplicate of the parent is recreated. The parent and child are exact replicas of each other ignoring the age difference like identical twins separated in time. Furthermore all cells would virtually be interchangeable between the parent and the time-clone.

Unfortunately, there is a natural penalty for fooling mother nature by wide scale cloning. The cloning process would never produce new individuals or new DNA. Cloning produces invariable duplicates. Normal fertilization mixes half the male chromosomes with half the female chromosomes to produce new and unique DNA. So, in this sense, the cloning process is a biological dead end.

On the other hand, cloning "desirable", bioenergy would probably result in more progress in certain areas than would ordinarily result from the natural gene pool.

Remember that bioenergy exists and behaves, while non-bioenergy only exists. Behavior is more than motion. The moon moves according to forces upon it which are produced by the gravitational fields of the earth and sun. The moon has no choice. However, living systems behave because of internal forces. Non-living systems have no choice about their own state of motion. Let us now turn our attention to how bioenergy behaves. What does behavior mean?

Chapter 4
Behavior of Bioenergy

Three Processes of Behavior

The three processes of conscious behavior are perception, decision and action (PDA).

Figure 8 shows a block diagram of these processes.

Unity of Perception

It is taught that there are five senses. These include the sense of touch. hearing, seeing, smelling and tasting. The truth of the matter is that there is a single common sense. It is the sense of touch. All five senses represent various refinements of the sense of touch. When one hears sound, molecules of air hit and bend the eardrum. When one sees, particles of light (photons) collide into the back part of the eye known as the retina. When one smells,

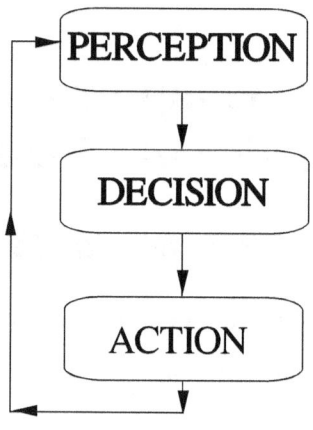

Figure 8   Three Processes of Behavior

molecules hit the olfactory nerves of the nose. Finally, when one tastes, molecules have come in contact with the taste buds of the tongue. In all cases the external energy has come in and made physical contact (touched us), generated an electrical signal and as a consequence, we felt it.

The point is that all bioenergy perceives its environment only through direct physical contact with it. This means that all experiences and original knowledge are based on physical contact. Systems of bioenergy can only perceive the world through direct physical interactive contact with the environment.

Bioenergy basically detects electrical signals produced by direct physical contact with the environment. Receptor neurons transmit electrical signals via the bioenergy's electromagnetic field of the body spirit so that the induced electromagnetic field of bioenergy's soul may interpret that detection. Again, it is the electromagnetic soul that perceives. The electromagnetic body spirit keeps the living molecules together. The soul controls the body spirit.

Decision Process

The decision process by the soul is the formulation of the next series of immediate behavioral commands based on the external perceptual inputs as well as internal inputs including past experiences. These behavioral commands control the physical action or chemical reaction of all the subsystems involved in whatever action the organism will perform next. The variations and number of behavioral commands are generally dependent on the type of bioenergy considered. The common basis for all decision processes is the satisfaction of what will be called the generalized drive.

The Generalized Drive

The generalized drive is the basic force which allows bioenergy to convert energy into itself. In all systems of bioenergy, the decision process is ultimately dependent upon a hierarchy of needs. In order of importance to survival these are:

(1) The desire/need for gas (oxygen for animals, carbon dioxide for plants)

(2) The desire/need for water

(3) The desire/need for food

(4) The desire/need for reproduction

The series of behavioral commands are formulated on the basis of the perceptual inputs influenced by the generalized drive and memory of similar past experiences. The generalized drive is hardwired into the DNA molecule of whatever the type of bioenergy is being considered. These drives form the primal directives for all bioenergy.

Action Output

It is action and only action that determines the influence of bioenergy on its environment. All actions are determined by the output of the behavioral commands originally formulated in the decision process. Operationally, action involves electrical signals sent by the soul via the body spirit while perception involves electrical signals sent to the soul from the body spirit. The soul is the controller of the body spirit which connects all of the bioenergy's molecular structure (body) and causes it to behave. It is only through action that bioenergy converts non-useful (non-consumable) forms of energy into (consumable) useful forms. It is only through action that bioenergy is capable of altering the environment for its own preservation.

Chapter 5
Human Bioenergy

Human Energy Form

In the not too distant past, planet earth probably contained many sets of species which were very similar to modern man. It would not be hard to imagine that at a certain place and time, out of the clay of the human like species which then existed on planet earth, did the first human being develop and emerge.

The human form of bioenergy is unique in the sense that it holds the highest rank of any species on the planet. No other species has the power to alter its environment to the extent that the human species can. We hold the power to control any other plant or animal on earth. In fact, humanity now contains the power to destroy all life on the planet. Ironically, humanity today cannot ensure its own long term survival and has not yet initiated any overall plan to do so. Human bioenergy has reached a stage where it is possible to study, describe and predict its own long term needs.

The human brain is indeed the most remarkable and complex nervous system on our planet. The experts all agree that the reason for this was the extreme survival pressure put upon our ancestors. The human body lacked the "tooth and nail" that most other species took for granted. Our physical characteristics alone could not turn away our predators. This precocious "do or die" situation multiplied by perhaps millions of years made it favorable for a brain capable of superior prediction to survive and flourish. It must be mentioned that all the life forms are related through the interaction between each of the electromagnetic souls and body spirits of all life forms. Since the human form has the superior brain, we automatically have inherited the destiny of all our plant and animal friends.

Life's Support System

The life sustaining force of this planet comes from the sun approximately 93 million miles away. Electromagnetic radiation from the sun is what supports our planet's bioenergy. Even though modern man supplements the sun's power source, mostly with fossil fuel, it was the sun which originally caused the fossil fuel deposits. One may think of fossil fuel as stored sunlight

Danger of the Fossil Fuel Supplement

Today, the people of the earth are partially sustained by the consumption of fossil fuels. These include natural gas, oil and coal. Most of the developed countries are dependent on fossil fuel for transportation, heating and electrical power generation. Moreover, most plastics and many medical products are derived from fossil fuel. The problem is that fossil fuels are not being replenished. This means that as time goes on, fossil fuel supplies are growing smaller and smaller. Thus, with the growing demand for fossil fuel, the price of fossil fuel will get more expensive over the long haul. Therefore, in the future, fossil fuel will have to be replaced or many people will perish.

Solution to the Fossil Fuel Problem

The only single technology which has the potential for replacing fossil fuel is nuclear power. This can be done safely and can be a two step parallel process. First, internal combustion engines will have to be designed and/or converted to burn hydrogen. Second, giant clusters of breeder (generating more fuel than they consume) reactors will have to be built underground and far away from population centers along with their radioactive waste and post waste processing facilities. The main purpose for these reactors would be to generate electricity used to convert supplied water into oxygen and hydrogen. The hydrogen could then be transported or even exported as fuel for use by the hydrogen fueled internal combustion engines. Note that this would help solve the global effects of burning fossil fuel since burning hydrogen produces pure water. Radioactive wastes would not have to be transported and would remain within the confines of the clusters. These nuclear clusters could also be used to desalinate and/or distill polluted water and would automatically return pure water back into the environment.

Future fusion power plants are the only alternative to present day fission reactors. Nuclear fusion is the only power source which offers a potentially vast, and relatively clean source.

The Greatest Battle

Bioenergy may be special in the sense that it is self organized and intelligent, and yet the human brain has not learned how to escape the great transition from the living to the non-living. Both energy forms must be accounted for in an assessment of the total. Thus, the total energy of the universe is composed of all bioenergy plus all non-bioenergy. This includes all the earth's bioenergy as well as any which may exist elsewhere.

This concept leads to a most interesting consequence. Since the total universal energy must remain constant, any additional bioenergy which appears (is born) must be accompanied by the disappearance of an equivalent amount of non-bioenergy. Conversely, more non-bioenergy ultimately means less bioenergy. There is one more important consideration. It takes energy to convert non-bioenergy to bioenergy. However, bioenergy requires no energy supply to convert to non-bioenergy (death). Since bioenergy represents a more organized form of energy than non-bioenergy, it must be continuously supplied by an external energy supply. This defines the seemingly unending struggle for survival. This is the greatest battle whose outcome is still far from being known. It is the battle between the forces of life and the forces of death. It is an ongoing battle for bioenergy to control enough non-bioenergy to guarantee its survival.

If bioenergy succeeds in bringing non-bioenergy under its control for the preservation of life and the ability to behave without the struggle to survive, then bioenergy will have won. If all bioenergy is eventually converted to non-bioenergy, then bioenergy will have lost. Ultimately, it is the action taken by bioenergy which will determine whether it will perish or survive. The action part of the perception, decision, action cycle is what contains the potential to rearrange energy for survival. Action is the result of decision. Decision is the result of perception and memory.

A Living God

Consider the following philosophically debated question. If God exists, then is God alive? The obvious answer is affirmative. Since bioenergy is known to exist on planet earth and since God represents the total invariable universal energy, and since the bioenergy of earth is part of the total, then it must follow that God is alive with measurable proof on planet earth.

Human Image of God

Recall that the Universal Spirit of God (all the field energy including God's Soul and Holy Spirit) serves to connect all the material parts of the universe since it is composed of all the force fields of nature. In comparison, human bioenergy is composed of material parts held together by the human body spirit controlled by the human soul. So just as all material parts of the universe are connected by the Universal Spirit, the human body is connected by the body spirit. Just as the human has a body spirit and soul, God contains the Holy Spirit and His Soul. These examples are at least two ways of how humans possess the "likeness of God".

Before creation and the big bang, there was no material energy. Only the induced electromagnetic fields (Soul and Holy Spirit of God) could have existed, as there was no charge or the motion of charge since charge always resides on matter. Now (after the big bang), part of God's Holy Spirit became the souls of the living. The Soul of God remains the same both before and after the big bang. Since any electromagnetic field is connected and influenced by all electromagnetic fields, it follows that not only is each human soul connected to the human body via the human body spirit, but each human soul is connected to God's Soul via the Holy Spirit. Note that there is always a living energy trinity associated with any human being, namely, the Soul of God, the soul of that human being and the associated connective energy of the Holy Spirit.

Even though each human contains a unique set of molecules, it is only through the controlling soul and connective body spirit that they all become dynamically integrated. It is this induced electromagnetic controlling soul which makes the body machinery alive.

The Unity of the Personal Living God

Recall the definition of God to be the total energy of the Universe. This total may be conceptually divided into the energy of any given human soul plus all the rest of the universal energy which will always contain the living Soul of God as well as all other living souls.

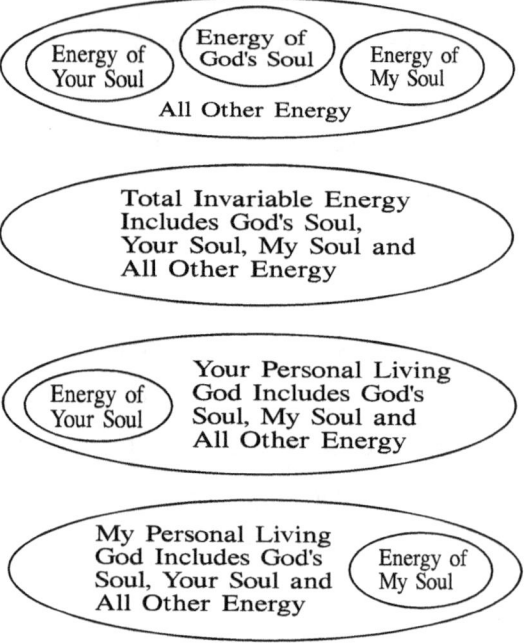

Figure 9   Personal Living God

Therefore, **the total energy (God) less the given human soul's energy (including the human soul's connective part of the Holy Spirit) is defined to be the personal living God for that**

human being. Clearly, this defines the induced electromagnetic field energy interaction between any human soul and all the rest of the universal energy. Since all the rest of the universal energy always includes the Soul of God (as it existed after the big bang), the electromagnetic interaction between any human soul and the Soul of God is also defined. Thus, even though every human soul is different, it has an induced electromagnetic connection to the identically same living Soul of God via the connective Holy Spirit.

Figure 9 demonstrates that my personal living God and your personal living God contains the same single Soul of God. Note that the total energy always remains the same in all four depictions, as God is an invariable (time independent) constant containing the single Soul of God.

Christian Trinity of Total Energy

God (the total energy) is equivalent to the energy of any living entity's soul, the energy of its personal living God, and the energy of its connective part of the Holy Spirit.

If the living being in question is Jesus Christ, then **the energy of Jesus Christ's Personal Living God (as The Father) plus the energy of the Living Soul of Jesus Christ, (as the Son) and the energy of Jesus Christ's connective part of the Holy Spirit is the same as the invariable God (the total energy)**. This corresponds to the idea in the Christian Trinity that God is The Father, The Son and The Holy Spirit.

Moreover, if the living being is you, then the energy of your personal God, the energy of you and your connective energy part of the Holy Spirit is also the same as the invariable God (the total energy). It can be demonstrated that both your personal living God and my personal living God contain the same living Soul of God as The Father. Please refer to figure 9.

Chapter 6
Human Behavior

Human Instruction Set

Whenever the human egg and sperm cell unite, the basic DNA hardware of the new individual is laid out. This hardware includes the static electromagnetic field which hold the DNA intact which is the beginning of a new body spirit. Each half cell contains the genetic code unique to its donor. The associated induced electromagnetic field energy received from both egg and sperm transforms into a new soul formed to control the new cell hardware.

The dynamic new cell that is formed contains a unique set of genetic code and electromagnetic soul which will control the one celled body spirit. This code contains all the information which will dictate the physical characteristics of the resulting new human. Each cell composing the human body contains the exact same set of genetic code as that of the first cell formed from these two living parental half cells.

Human Nervous System

A most interesting subset of human cells is called the nervous system. This is composed of specialized cells called neurons. These neurons are capable of conducting and generating electrical current. The nervous system is the vehicle by which the electromagnetic soul controls the human body spirit.

The spinal cord is the electric cable which connects all the peripheral organs and muscles with the electromagnetic soul mainly seated within the human brain (an organ composed of neurons). The soul constantly surveys and monitors (via electrical signals) all of its peripheral hardware. It perceives the signals coming into the brain. It controls the signals going out from the brain to the organs and muscles for the accomplishment of the desired action or chemical reaction. As the action takes place the soul monitors the actions and compares it

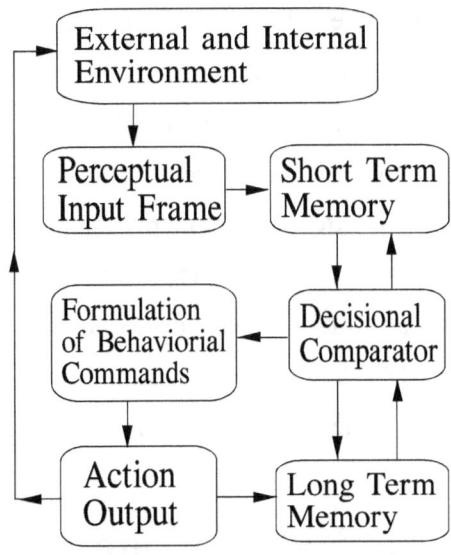

Figure 10   Detailed Processes of Behavior

with the desired action. Nevertheless, between the inputs to the brain and the outputs from the brain is the decision process made by the induced electromagnetic soul.

A Model of the Human Mind

Figure 10 is a block diagram of detailed behavior. External means external to the human body. Internal means inside the human body. This is what is meant by the top block in the figure, called External and Internal Environment. In other words, all signals either arrive externally or internally. Note that the soul is in direct electrical internal contact with the body spirit The body spirit channels both external and internal stimulus to the soul. It is the electromagnetic human soul that receives perception, makes decisions and causes actions (PDA). It is the human soul that directs the spirit to cause the body to behave.

External and Internal Environment

The body spirit channels signals to the soul from both the external (outside the body) and internal (inside the body) environment. This means that if the human body changes (such as form, position or chemistry), then the internal environment also changes. The external environment continuously puts out a spectrum of energy which include particles of light, sound waves, atoms, and molecules. The body spirit also internally produces a stream of electrical feedback signals to the soul.

The sum of all external energy particles from the environment that impact the body are transformed into electrical signals by the associated "external perceptual neurons". Both the internal body signals as well as external environmental signals form an "input snapshot" perceptual input frame. The information contained in this perceptual frame is then channeled into the short term memory of the brain (external environment via optic nerve for eyes, auditory nerve for ears, etc. and internal signals via internal perceptual neurons).

"Processing" neurons are then directed to make a comparison between short term memory and long term memory. One may think of long term memory as the store house of all past experiences for a particular human life. Long term memory contains all past "input perceptual frames" as well as the corresponding successful "behavioral commands" which were issued in response to the past input frames. Thus, the processing neurons form behavioral commands by a systematic comparison between "present input frames in short term memory" and "the permanent past input frames in long term memory". The resulting behavioral commands reflect a maximum compatibility between short and long term memory. At the same time, these behavioral commands are also modified (refined) on the basis of the difference between predicted and actual satisfaction of the generalized drive.

The behavioral commands are often referred to as motor commands because as the commands are executed, the result is electrical signals being sent to the appropriate muscle cells or other internal organs. This action output is carried out via the effector neurons under the jurisdiction of the electromagnetic body spirit controlled by the soul. Once the muscle cells or organs have been stimulated electrically, the result is a change in the internal (body) and/or external environment which begins the PDA cycle over again.

Long term memory will sometimes be referred to as "self image" memory. Others may refer to it as the "subconscious" mind or as the unconscious mind. Nevertheless, one may readily see that it is the long term memory which influences the action outputs which serve to alter the internal and external environment.

Consciousness and Unconsciousness

Consider the definition of consciousness as **the real time or present time cyclic process of perception, decision and resultant action.**

Conversely, **unconsciousness** or the dream state (as opposed to an injury or disease induced coma) **would be the cyclic process of imagined perception, decision and resultant imagined action**. Such PDA cycles result in the programming of long term memory. Note also, that this may sometimes lead to an action output such as rapid eye movement (REM) or other muscle/organ stimulation. Note that imagined behavior (PDA cycles) do not occur in response to real or present time perception.

Normal REM sleep is the process of purging short term memory and enhancing long term memory in preparation for the next conscious waking cycle. The successful PDA cycles in short term memory reinforce the self image PDA cycles in long term memory. Thus, REM sleep is seen to be essential for good mental health by preserving the consistent relationships between long and short term memory. Note that long term memory represents the true self image standard when purging the PDA cycles in short term memory.

Thus, there can be several natural states of unconsciousness such as sleep, hypnosis, meditation or prayer. While sleep is seen to be a cleansing process for the mind, other forms of unconsciousness may serve to reprogram long term or self image memory. Careful sessions of hypnosis, meditation or prayer may serve to heal both the mind and body by utilizing vivid imagination of "desired" behavior. This process is presented in chapter 8 under the modification of behavior section.

Chapter 7
Human Brain

Parts and Function

L et us now examine the main parts and
functions of the human brain. Many have
compared the brain to a computer. Indeed,
the human soul may be thought of as the
programmer, or controller of the human brain.

Corpus Collossum

The physical characteristics of the human brain
include what is termed the corpus collossum. This
portion of the brain separates the left half of the

brain (the left hemisphere) from the right hemisphere.

It turns out that the left half of the brain is associated with the right portion of the body.

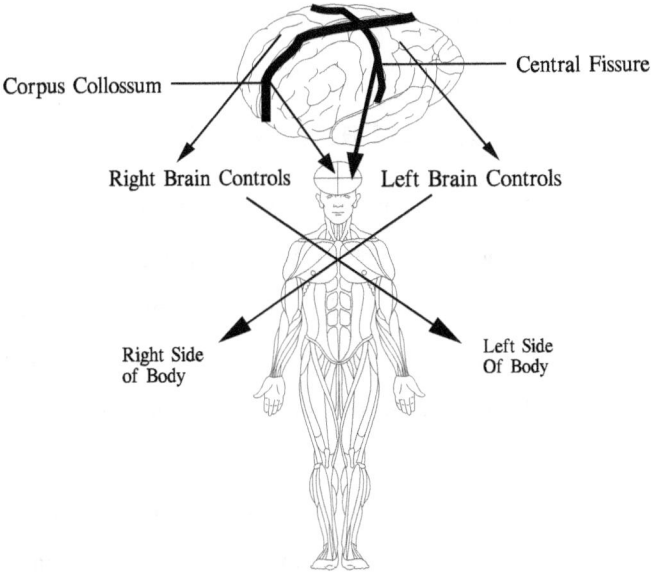

Figure 11   The Corpus Collossum and Central Fissure

Additionally, the right half of the brain is associated with the left part of the body. These relationships are shown in Figure 11. The brain is slightly rotated to show its main fissures.

Recent brain research indicates that generally, the right hemisphere is associated with emotions while the left hemisphere is associated with logical thinking. The left hemisphere usually contain the person's speech center. That is, the left hemisphere is responsible for both word recognition and speech.

Central Fissure

Another major characteristic of the human brain is called the central fissure. This portion of the human brain separates the front part of the brain from the rear portion of the brain. It turns out that the front portion of the brain is responsible for "motor control". Motor control refers to the electrical signals which are directed from the brain (via the electromagnetic soul) to the particular muscles and/or organs of the body (via the electromagnetic body spirit). Moreover, the rear portion of the brain is responsible for perception. This perceptual portion of the brain is what receives the electrical signals which resulted when the environment (both internal and external) impacted the body. Between the perceptual and motor control is the decisional portion of the brain.

How You Drink From a Glass of Water

It may seem trivial to explain something as simple as drinking water from a glass, but what the researcher finds is that there is no medical, biological, psychiatric or analytically acceptable explanation of how a human being drinks from a glass of water.

However, armed with the tools and facts presented in this chapter and the previous chapter on human behavior, the scientific explanation is as follows:

(1) In a previous perceptual input frame, the soul had perceived that there is a glass of water within reaching distance and that the body can use and hold a drink of water. This is the perception part.

(2) The soul decides that the body should drink water. This is the decision part and was based on past similar experiences.

(3) The soul then initiates all the motor control programs (actions stored in long term memory) that the body spirit must perform in order to take a drink.

This includes the electrical signals sent to and feedback from the visual cortex and various muscles of the arm and fingers to reach and grasp the glass, lift the glass to the mouth, the muscles of the jaw, tongue and throat to sip and swallow the water, feedback from the throat and stomach, and finally, for the arm and fingers to set the glass back down. This is the action part.

Dual Redundancy of Perception and Action

The perception of the glass of water in step 1 resides in short term memory behind the central fissure. Perceptual input frames come in from the left and right half of the body. The back left hemispherical portion of the brain receives incoming signals from the right portion of the body. The back right hemispherical portion of the brain receives incoming signals from the left portion of the body.

If the human used the right hand, all the programs and routines for executing step 3, reside in the front left hemisphere of long term memory. These motor control programs in long term memory have been learned and refined based on the past successes of drinking from a glass. Note that these programs are duplicated in the right front hemisphere of long term memory since one could just as well used the left hand to drink.

Thus, there exists a dual redundancy (left and right hemispheres of the brain) for both perception (signals coming into short term memory) and action (signals going out) from the motor control programs in the long term memory of the human brain.

Chapter 8
Human Soul and Spirit

The total electromagnetic field of the human being may be divided into a controlling portion (the induced electromagnetic soul) and the body spirit (electromagnetic fields associated with the body). The body spirit keeps the body together. The human soul that controls the body spirit is also responsible for directing the function of the human brain.

Recall that magnetic fields arise from either charge in motion or time varying electric fields. Electric fields arise from the presence of charge or time varying magnetic fields. Electric fields which results from time varying magnetic fields or magnetic fields which results from time varying electric fields are called induced fields. Again, note that these induced electromagnetic fields of the human soul exist apart from the material universe since they are neither caused by the presence of charge or the motion of charge.

There is indirect evidence that this human soul may either enter or exit its human body. Note that if the human soul permanently exits the body, it also relinquishes control over the body spirit which can temporarily hold the material body together.

Spiritual Survival of Death

Dr. Raymond Moody in a book entitled "Life After Life" presents some indirect evidence which suggests that the soul is capable of surviving death. In fact, after interviewing several people which had been pronounced clinically dead, he was able to propose a model which contained a general scheme which described the death experience. The model included the "out of body" experience while dying; the events which occurred and were experienced by the soul while out of the body; and further, the return of the soul back into the body. This model reinforces the possibility that the electromagnetic soul may survive death. It is only through further investigation and experiment which will eventually answer whether or not the soul can survive death. Recall also, that the energy of each individual soul cannot be created nor destroyed.

Modification of Behavior

As one examines detailed human behavior, it becomes more and more apparent that it is long term memory which governs a person's behavior. Recall that short term memory is responsible for conscious memory and long term memory makes up the subconscious memory. It is estimated that long term memory represents about 80% of total memory while short term memory is roughly 20% of the total. Indeed the power of the mind lies in the subconscious.

It must be emphasized that long term memory trends of a typical human is basically programmed by the age of twelve years. The decisions by the soul are thus guided by what is in long term memory. All action outputs must be formulated to be consistent with long term memory. The individual literally cannot behave against long term memory. Long term memory represents absolute truth to the individual. Only with difficulty, can it be altered by the conscious soul. Long term memory represent what is possible for the individual to achieve or even attempt. It represents how a person feels in his heart about himself. It represents how a person feels about life. It controls whether a person is positively oriented or negatively oriented. It literally dictates what one will even attempt to accomplish. Everything in long term memory is assumed to be error free (even if it possesses obvious errors such as the long term memory of dangerous criminals). After all, it has passed the tests of nature, survival and satisfaction of the generalized drive.

A person's habits or behavioral patterns can only be changed by altering the programming in long term memory. Once long term memory is altered, then behavior is changed. One way to alter behavior is through vivid imagination. Long term memory cannot distinguish between real behavior and vividly imagined behavior. If one pictures oneself doing what is desired, then long term memory can be altered. This is called simulated behavior. Figure 12 is a block diagram of simulated behavior.

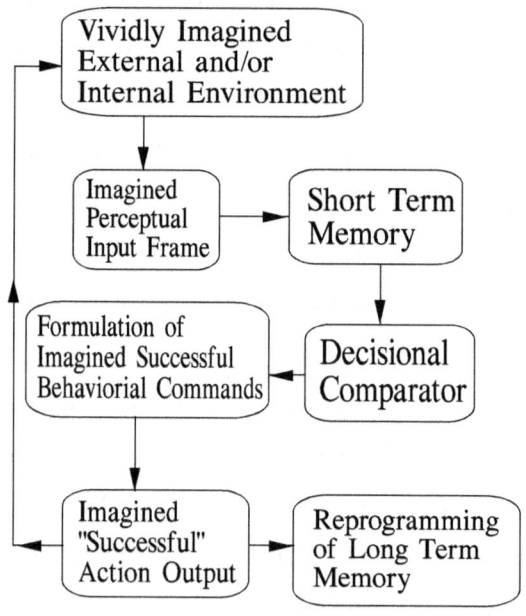

Figure 12   Simulated Behavior

This may be compared to Figure 10 which shows a block diagram of detailed behavior. The main difference between actual detailed behavior and simulated behavior is that the decisional comparator (soul) does not consult long term memory in real time. It formulates imagined successful behavior from short term memory alone. Once, the imagined behavior is performed, long term memory records this as a successful action reprogramming. In a book entitled, "Psycho-Cybernetics", Maxwell Maltz describes how desired behavior is achieved through the use of imagination. Simulated behavior gives the detailed mechanics behind the reprogramming of self image (long term) memory. Thus, there are ways to reprogram the unconscious mind, so that there can be genuine rehabilitation for the mentally ill mind as well as potential improvement for all desired real time behavior.

Sickness, Health and Long Term Memory

If one's self image is a picture of frailty or sickness, then one will follow the behavioral patterns of sickness. Recall that self image represents truth and so if it contains a sick self image, then that behavior must follow and one will be sick. On the other hand, if one's self image is a picture of health, then one's behavior will automatically follow the path to health.

The best mental exercise to do if one is injured or gets sick, is to vividly picture oneself to be well. This will prepare both body and mind to accept wellness as truth and thus, facilitate the natural healing process. In many cases, pain may be used as a trigger to begin this wellness exercise. As an example, say, if one has a broken arm in a cast, then the occasional pain from the broken bone may be used to trigger one to vividly imagine that arm working perfectly well and healthy.

Chapter 9
Knowledge

Questions of Knowledge

To have faith that something is true and to "know" that something is true are two different states of affairs. We may believe that if a pencil is dropped, it will fall to the floor, but we "know" that it fell to the floor only after it is dropped and indeed falls to the floor. If the pencil was dropped many times, and each time it fell to the floor, then on this basis one could "know" it would fall to the floor if dropped. What is knowledge? How do you know that what you believe is true? Are there truths and knowledge beyond the human interpretation?

Knowledge and God

All knowledge is revealed through experimentation. Since all experimentation involves energy transformation and since energy is only transformed by the dictates of the total force fields of nature, all measurements are essentially spiritual in nature. Recall that there are only four fundamental force fields of nature. The electric and magnetic fields affect charged particles, while gravitational and nuclear fields affect all particles. Recall also that the total force fields of nature represent the Universal Spirit of God. Thus, knowledge is revealed by the questions to God (the experiment) and answers from God (the experimental result).

Two kinds of Knowledge

Human knowledge is of two kinds. Descriptive knowledge allows for the description of past and present events. Predictive knowledge allows for the prediction of what will happen in the future. Both types of knowledge essentially allow for self preservation. Bioenergy tends to accumulate only the knowledge which allows it to survive. The human fund of knowledge represents a considerable extension to the basic survival instinct. Bioenergy inherits a certain amount of basic DNA knowledge through the genetic code. It is through the electromagnetic fields of bioenergy that this knowledge is passed on from generation to generation.

Experimental Result

Experimentation is the basis for all knowledge other than the inherited DNA hardwired genetic material. Experimental facts support descriptive knowledge. The integration and organization of a large body of experimental facts support scientific theories. Scientific theory allows also for the "prediction" of new experimental truths. The more experimentation which is carried out, the more that humanity knows about the behavior of energy. Human knowledge which allows for more efficient survival is termed survival value. More survival value means more control of humanity over its environment.

Nature of Physical Reality

Experimental science does not concern itself with all the various philosophical definitions of reality. It bypasses this consideration entirely by assuming that only those quantities which lend themselves to measurement (through experiment, experience or observation) are real. Things that are imagined to exist may only be validated or invalidated through experiment. Things which are claimed to exist without the possibility of observation or experience are meaningless. Of course, this is not to say that one should not investigate what one intuitively feels is truth. Investigation implies experimentation. The ideas of reality may be different than reality itself. It is only through measurement which allows one to distinguish between what is real and what is thought to be real. Measurement is what distinguishes between truth and falsity. To say that what is measured is an apparition and that "something else behind the scenes" is real, leads to a self contradiction. One could always claim that "the something else behind the scenes" is also an apparition. Thus, only what is or can be measured is real and exists. All which has the potential to be measured forms the "true" reality.

Knowledge and Reality

All branches of science seek a common goal. Fundamentally, the goal is to describe and predict nature. As previously mentioned, experimental facts form the foundation of established scientific theory. Experiments may be carried out to prove or disprove theory. If a theory is put forth that cannot be experimentally tested, it adds nothing to human knowledge and is mere speculation. There is a great difference between something which contains the potential to be measured and ideas which do not. It must be mentioned that many experiments may reinforce scientific theory, but it only takes one experiment to disprove it.

Two Results of Knowledge

The result of knowledge lies in its application. Many of the great experimental discoveries have had positive as well as negative results. The application of knowledge which adds survival value to bioenergy is a positive result. Applied knowledge which takes away survival value is a negative result. The application of nuclear technology towards mass destruction and disorganization of bioenergy is negative. Application of such technology towards providing a safe, reliable source of power is positive. Knowledge itself is neither good or evil. It is the application of knowledge which is either positive or negative and subsequently good or evil.

Knowledge and Survival

The struggle of survival began simultaneously with the appearance of bioenergy on planet earth. This was the beginning of the great battle which has furiously raged into the present time. Survival means the satisfaction of the generalized drive for all the subsystems of bioenergy. Thus, an adequate supply of fuel must exist to support whatever biosystem is being considered. For most all bioenergy, the sun has furnished the fuel. All systems of bioenergy require food supplied by other particular biosystems. This supply and demand define the cooperation which exists for all life forms. Disturb one link in this biosystem, and the entire chain is disturbed. Natural checks and balances have always existed within the biochain to insure the integrity and security of each link. On the low end of the spectrum, there are individual single celled plants and animals. On the high end, are the multi-celled biosystems. Each is capable of altering its environment. Most of Earth's life forms have very little potential to alter their environment for the purpose of securing their own preservation. On the other hand, the appearance of the human species has changed the old picture.

Knowledge translates to survival power. This also refers to the predictive type of knowledge (knowledge that allows one to predict what will happen before it happens). Recall that successful predictions guarantee survival via the satisfaction of the generalized drive.

On the other hand, the lack of certain knowledge represents the constant threat to the survival of earth's bioenergy. It is the attainment of certain knowledge that can guarantee the survival of what remains of earth's life forms. It is also the lack of certain knowledge which may guarantee the destruction of earth's bioenergy.

Chapter 10
Scientific Laws of God

Natural Law

All the laws of nature have been discovered and formulated by the process of experimentation. Experimentation is the careful observation under controlled conditions, of how nature behaves. Experimentation reveals facts of nature. These facts of nature give rise to scientific theory which embody one or more of the laws. These laws represent relationships between the fundamental universal descriptors. Human bioenergy has extended its knowledge base by current experimentation in the hope to discover all the laws of nature. It is these laws which yield the ability to describe and predict. Questions to God are submitted in the form of an experiment. God's answers to these questions are given by the experimental result. Our ideas about how things work are either proven or disproven through experiment.

God's scientific laws are fundamentally either mechanical or electromechanical. This is because energy is either uncharged or charged. Mechanical force fields are either nuclear or gravitational. Electromechanical force fields are either electric or magnetic.

The scientific laws of God that have been uncovered through experiment represent man's fund of knowledge. As more and more experimental facts are gathered, God's scientific laws become more accurately revealed. So it is that the laws of nature find mathematical quantitative expression. Either Newton's basic law of gravitation or Einstein's general theory of relativity allow us to predict gravitational fields in space very accurately. Maxwell's equations allow us to predict both electric and magnetic fields. So what basic laws are missing? One has to do with bioenergy and the other deals with non-bioenergy.

Unknown Laws of the Cell Nucleus

The first weakness in our fund of knowledge is that the dynamical laws governing the basic constituents of bioenergy remains undescribed and therefore unpredicted. On a microscopic scale this means that bioenergetic DNA behavior eludes accurate description. Thus, the dynamics of the cell (the fundamental unit of all bioenergy) is not fully understood. The cell is controlled by the cell nucleus. For multi-celled bioenergy, the cell nuclei contain the DNA molecules. Thus, if the dynamics of DNA molecules were known, then essentially the cell dynamics could be known. The eradication of disease will only be possible through more knowledge of the dynamics of the cell. The ability to describe the dynamics of a cell will eventually lead to the ability to describe large groups of cells.

If cells could be grown and controlled, then there would be a steady supply of healthy cells to replace destroyed cells. Both cell performance and cell health could be guaranteed if enough were known about how the cell behaves on a molecular level. This would also explain the behavior of nerve cells.

Unknown Laws
of the Atomic Nucleus

The second great weakness of human knowledge deals with non-bioenergy. It has to do with the dynamical laws governing the basic constituents of atomic nuclei. Atomic nuclei are composed of nucleons (neutrons and protons). Even though, the standard model identified the bosons (pions) which are responsible for the forces between nucleons, the underlying dynamics of this force is not completely understood. Thus, this force remains undescribed and the vast storehouse of nuclear transformation energy remains untapped. In other words, if these nuclear force fields were understood from a "dynamic principle" point of view, thermonuclear fuel could be mass produced. Controlled thermonuclear power plants could be built and the fuel shortage problem (which all of us will soon have to face) would literally disappear. With a vast supply of cheap and relatively clean thermonuclear power, almost anything could be grown indoors and anything could be built (including long range spacecraft). The fossil fuel supply could be spared, large scale pollution and global effects of burning fossil fuel could be curtailed and the solar system could be colonized. The richness and value of any society can be measured by how much fuel is at its disposal. Everyone would literally be rich on a relatively clean thermonuclear power source.

Knowledge and Scientific Law

Both physical and spiritual laws inherently contain descriptive and predictive knowledge. They also express the accuracy of such description and prediction. In a sense, the natural laws of God reveal to us our limitations. They indicate both what can be accomplished (in the sense of energy transformations) and what can be known (in the sense of experimental measurement).

The basic laws of physics which are included in this manuscript, are divided into three major sections. These are mechanical laws, electromechanical laws and conservation laws. Mechanical laws include Newton's dynamics, Einstein's special relativity dynamics, thermodynamics and quantum mechanics. Electromechanical laws include Maxwell's equations which are the basis of optics, the Lorentz force and the induced electromagnetic wave equation. The conservation laws express all the quantities which remain the same both before and after energy is transformed.

Chapter 11
Morality

Common Elements
of Ethical Philosophy

When one examines the various moralistic viewpoints, some common elements may be noted. "Good" behavior usually brings about pleasure, virtue or happiness. Moreover, pain and unhappiness are presumed to be the result of bad behavior or immoral behavior. The greatest problem in all ethical philosophy is that terms such as good, bad, evil and virtue are for the most part left undefined or are defined in an inconsistent manner. Correct or "right" behavior as well as incorrect or "wrong" behavior can only be recognized if by what is meant by such terms are clearly and unambiguously made comprehensible. These definitions should be relatable to a basis, which has meaning with respect to all systems of life. The true principles of morality should be based on the survivability of all bioenergy.

Behavior and Morality

Remember that morality implies behavior, and that all bioenergetic behavior consists of perception, decision and action. In the world of bioenergy, there exists struggle in the sense that each species competes with the next, and in the sense that a particular biosystem represents food to another. We are presently the major cause of the earth's pollution. Our pollution is a side effect of our hunger for and consumption of the earth's resources. Our nuclear weapons are capable of wiping out life on earth. If the human race is to survive, it must accept a sound basis of morality.

Basis of Morality

Essentially **the basis of morality is the preservation and survival of all the life forms in the universe**. If a behavioral act goes along with the survival of bioenergy, then it is moral. Directive survival is behavior which preserves and enhances the total bioenergy. Directive survival for the individual means the satisfaction of the generalized drive. Conversely, if an act goes against survival, it is immoral. Note that an individual could behave for self preservation and at the same time go against the preservation of the group. The total biosystem is more affected by a given species than any one member of that species. Thus, if an individual performs self survival actions which go against the survival of the species, then that behavior is immoral. On the other hand, the rules of law and social behavior must maximize the freedom and survivability of the individual.

Extinction of a Species

Bioenergy is observed to take on many forms. The natural law of supply and demand guarantees that all biosystems are related. Each biosystem depends on one or more of the other biosystems. Each biosystem consumes others and in turn gets consumed. Thus, there exists a natural balance throughout the bioworld which has more or less preserved each species which has survived. If a species becomes extinct, all other species may become affected. Moreover, if a species becomes extinct, it most likely cannot be revived in the sense that its genetic code may be forever lost. The major unbalanced force which is presently being placed on the total biosystem is the proliferation and consumption by the human species. Some species may adapt to the demands placed on the environment by humanity and some may not. Ironically, the human species has reached a stage in its development where it is capable of destroying all other species including itself.

Test of Survival Value

To determine whether an action is moral or immoral, one must best predict the impact of such action on all other bioenergy. If the bioenergy involved gains survival value, then the action is moral and good. If the bioenergy looses survival value then the action is immoral. In the determination of whether an action is moral or immoral, the total bioenergy must be taken into account.

The survival of any particular species of bioenergy is dependent on the corresponding food chain. This food chain is also composed of bioenergy. Overpopulation is the other extreme of extinction. In determining whether an act is good or bad, it is also necessary to evaluate the relative strength of related population numbers. When a species overpopulates, individual members of the species lose importance and freedom. When a species is on the verge of extinction, the few individuals of that species become important in the sense that they alone, hold the key to their future generations and their link to the biochain.

Chapter 12
Survival of Bioenergy

A Question of Survival

As more and more stress is placed upon the entire biosystem, with the continued consumption and demand for food and power resources by humanity, one wonders what the outcome will be. Most all species that survive will eventually be controlled by the human species. Will bioenergy on planet earth preserve itself and survive or will it eventually perish from the earth? Remember, because of the definition of God, if bioenergy perishes from earth, then God is dead on earth. This question is even more serious when one considers that intelligent life beyond the earth has not yet been confirmed. The quest for other life has included exploring the solar system with spacecraft and exploring the stars with radio telescopes. Mankind has been seriously searching for intelligent life for the last half century with negative results. There is the distinct but minute possibility that life exists only on planet earth. For planet earth, the struggle to survive represents the final battle. It will either be won or lost. Moreover, life on planet earth may have been the only seed which could distribute life throughout the rest of the universe. It is the behavior of human bioenergy which will determine the outcome. Let us now examine what is known about our limitations.

Limits on Motion

Nothing (in the sense of energy) has ever been observed or measured to travel faster than the velocity of light in free space. Even though photons (conceptually as electromagnetic waves or quanta of light) possess mass in flight, they possess no measurable mass when they are at rest (do not move). As more and more experimental data became available, the laws expressing the motion of energy systems having rest mass became quantitatively expressible. These are now known as the laws of Einstein's special theory of relativity and are included in the fundamental physical laws towards the end of this manuscript. From these laws in accord with the revelation of knowledge, it was seen that systems of energy possessing finite rest mass could only approach the velocity of light in free space but could never arrive at or exceed it. Only the systems which contained no rest mass are allowed to move at this limiting velocity. All experiments to date have confirmed this conclusion. It must be emphasized that if only one experiment shows any energy system to travel faster than light in free space, the present laws will break down and a new scientific theory will be formulated that will represent a closer approximation to the exact scientific laws of God.

Limits on Knowledge

The Heisenberg uncertainty principle expresses a limit on what can be known in the ideal experiment. It expresses the fact that the momentum (mass times velocity) of an energy system and its position cannot be precisely known (measured) simultaneously. If the position of the system is precisely known, then its momentum cannot be determined. Conversely, if the momentum is known precisely, then its position is undetermined. Another form of this principle is that a system's total energy and the time that it had this energy cannot be known simultaneously. Note, that this uncertainty is built into the universe and does not result from inaccurate experimentation. This is knowledge that God removed when he changed Planck's constant from zero to a small positive value. This forever changed the laws of the universe from deterministic to probabilistic and caused a transformation of energy (before the big bang) into a living material universe (after the big bang). This uncertainty is now built into all systems in motion including living systems. It is now identified as "Free Will" and essentially prevents the living energy's behavior from being completely known.

If Planck's constant had been left at zero, there would be no uncertainty and the exact behavior of all living energy systems could be known. All bioenergetic behavior would be controlled by the Will of God and could be compared to living mindless robots. If our exact behavior cannot be known, then it cannot be controlled. Thus, we were given free will to behave for or against our own survival, as well as the survival of our species. Moreover, one is free to either accept or reject both the idea and existence of God.

Nature of the First Cause

In 2002, a theoretical discovery was made by the author in a study that uncovered the triggering mechanism behind the big bang. This study resulted in a scientific theory entitled "Nature of the First Cause". It was found that before the big bang, time fluctuations could oscillate between positive and negative values. A negative oscillation could produce negative antimatter, while a positive oscillation would produce positive matter. It is well known that positive matter gravitationally attracts other positive matter. It turns out that negative antimatter gravitationally repels positive matter and negative antimatter gravitationally attracts other negative antimatter as predicted by Newtonian as well as Einsteinian gravitational theory. First Cause theory assumes that the total material (mass and charge) energy (therefore including mechanical and electromechanical) was zero before the big bang. Note, that this number only accounts for material energy and would not include any self induced electromagnetic field energy before the big bang.

Thus, when a time fluctuation was quick enough and negative, it produced net negative antimatter and its associated positive repulsive gravitational field energy. When time fluctuated positive on its way back to zero, it produced positive matter which was then repelled by the repulsive gravitational force field energy produced by the negative antimatter. This positive gravitational field energy plus the positive mass energy precisely balanced the negative mass energy. This positive gravitational force field caused a large repulsive force between the negative antimatter and positive matter which caused them to accelerate away from each other. As shown by Einstein's special theory of relativity, as velocity increases, so does mass. Therefore, both the positive matter and negative antimatter gained mass magnitude as their mutual repulsive velocities approached the velocity of light. This in turn, caused more positive repulsive gravitational energy (caused by the increasing negative antimatter) and this runaway situation caused the universe to rapidly expand and gain both matter, positive gravitational field energy and negative antimatter. Part of this positive matter was destined to become planet earth and ourselves.

If only a small amount of positive matter and negative antimatter were originally separated by a very small distance, their separation force would have been quite violent and the first matter in motion would have been photons and negative anti-photons since they have the smallest rest mass (zero) and move at the fastest possible speed (velocity of light in vacuum). Thus, first cause theory predicts that the first matter in motion to appear were particles of light or electromagnetic radiation.

Negative energy anti-photons would be repelled by positive energy photons as well as the positive matter. Positive energy photons would be repelled by negative energy photons as well as negative antimatter. Thus, there would be a rapid accelerating spatially expanding demarcation between positive matter and negative antimatter (dark energy) and between light and dark (negative energy) photons. Furthermore, matter and negative antimatter both contain charges and charges in motion and therefore, electromagnetic field energy.

If the equivalent mass of this repulsive gravitational field energy is considered matter, then this first cause theory preserves matter-antimatter symmetry and solves the observed asymmetry between matter and antimatter. Moreover, this positive gravitational field energy potentially solves the missing dark matter problem of cosmology.

First cause theory also predicts the correct estimate for the number of stars (as observed and approximated by astronomers) in the universe. First cause theory also explains dark energy (negative antimatter) and why the galaxies are accelerating away from one another. Matter and negative antimatter are seen to form two entangled parallel worlds. Finally, first cause theory predicts that the entire universe behaves exactly as a black hole having the same dimension as the Schwarzschild radius as its event horizon. This horizon describes the leading edge of the universe defined by both an expanding gravitational field energy as well as the first escaping photons of the rapidly expanding matter and negative antimatter components of the big bang.

First Cause Theory Verifies Genesis Creation Account

Before creation, it has been shown that the only field energy that existed in a non-matter (uncreated) state would be induced electromagnetic field energy. Literally, this translates to a dark, formless, void (vacuum of space) containing the Universal Spirit as the dynamic Soul of God and His connective Holy Spirit.

First cause theory predicts that the first matter set in motion that was subjected to such a large creation force was the electromagnetic field energy associated with photons (particles of light) since they had the least inertia. Time began with the creation and motion of light.

First cause theory predicts that the initial void of space was quickly divided into two entangled parallel material worlds. One world is made of negative antimatter which gravitationally repels the other world made of positive matter. Both systems accelerate away from each other and quickly approach the limiting velocity of light. Both systems are invisible to each other except for their entanglement with an expanding repulsive gravitational field. Thus, darkness (negative antimatter including negative photons) was separated from lightness (positive matter and positive photons). This separation is literally between what can be seen (i.e. in the day) and what cannot be seen (at night).

Obviously, part of the positive matter energy became earth, sun, moon, other life forms and finally ourselves. All the rest of the positive matter energy became stars and other positive energy forms beyond our solar system. The negative system remains dark.

Thus, the basic biblical account of creation is analytically substantiated by first cause theory. Moreover, the total material energy of the universe is the antimatter energy, with the repulsive gravitational field energy, plus the matter energy must all add up to zero material energy. The total induced electromagnetic field energy must be the same now as it was before the big bang. Additionally, all antimatter energy and all matter energy are in accelerated motion with respect to each other and are being propelled via the Universal Spirit.

If the constant energy of the universe before the big bang was the electromagnetic energy of God's Soul and His connective Holy Spirit, then God's implementation of the final law of Heisenberg uncertainty is what caused the creation event. The uncertainty principle is the basis of the non-deterministic or probabilistic universe that now exists. It is also the basis of quantum mechanics, chemistry and the laws of life. The uncertainty principle gives living energy systems a certain freedom ("free will") from knowledge of their exact behavior. Thus, this law of uncertainty was the "first cause" which created the expansion of the total universal energy and produced all life forms. Before the big bang, God was a controlling Soul and a connective Holy Spirit. As such, God could know everything since there was no uncertainty. The universe before the big bang was completely deterministic. However, after the big bang (creation of the universe), the universe became a probabilistic quantum mechanical energy system in motion. The most that can be known about energy systems in motion are their wave functions. God's law of uncertainty implemented as free will before the instant of the big bang, turns out to be His gift of life. Thus, the first cause theory offers the first mathematical scientific creation theory.

Free Will

Free will is associated with the Heisenberg uncertainty principle because this law expresses the fact that not everything can be known about energy systems in motion. Since all life forms are energy systems in motion, it follows that their behavior cannot be exactly determined. This is precisely why all life forms have free will. This also means that since their exact behavior cannot be known, neither can their exact behavior be directly controlled. The Soul of God   (before the big bang) sacrificed exact knowledge of systems in motion by implementing the uncertainty law, so that in return (after the big bang) other (besides the Soul of God) free willed living entities would be created within the expanding universe.

Conversely, if life forms did not have free will, then it could be argued that all behavior and the motion of matter could be predicted and controlled by the will of God. Thus, the creation of an evolving universe would be meaningless, since the final state of the universe could be predicted and known with certainty. This would mean that all of God's living creatures (as robots controlled by Him) would have no choice in their own long term destiny. Moreover, the worship of God by his robots could have little meaning.

Quantum mechanics teaches that the most that can be known about energy systems in motion are their wave functions, but their exact position and motion cannot be known simultaneously. This is also the reason that life forms have free choice to either behave for survival or behave against survival. This in turn, means that any kind of (good or bad) behavior is caused by the controlling free willed soul within, and not necessarily the result of God's direct control. The decision process (one of the three components of behavior as in perception, decision and action) involves a choice between several possible good or bad actions. Without free will, the decision process would be meaningless because behavior would be reduced to perception and resulting action as in a table lookup pre-programmed robot.

The Balance

Clearly, for earth's bioenergy to survive, it must reach and maintain a balance between resources and population numbers. Bioenergy which is not directly controlled by humans, must be carefully preserved. Bioenergy which is food for humans must be carefully farmed, harvested and controlled by humans. Bioenergy consumed by humans which is not directly farmed and controlled, must be carefully studied, monitored and limited until methods or controls can be implemented so that survival of that species can be made more probable. Most of earth's oceans, lakes and rivers are being polluted and ravaged on a daily basis by both human waste and over fishing. Earth's oxygen producing forests are rapidly disappearing. Enforceable national and international agreements must be implemented in an attempt to preserve earth's remaining   plants, forests and water inhabitants.

The given solution to the fossil fuel problem outlined in the chapter on human bioenergy should be started. If the fission power plants and their waste sites are kept together, then many safety concerns could be minimized or even eliminated.

Experimentation must be accelerated, especially towards understanding the dynamics of the living cell and the dynamics of atomic nuclei. Once the nuclear force field is completely understood, a vast and relatively clean power source can be provided. Moreover, once the living cell is understood, both physical and mental disease can be controlled and eventually eradicated.

With plenty of power and health for all life forms, humanity will have reassumed its responsibility to enhance the preservation and proliferation of the livingness of God on earth and beyond.

GLOSSARY

ACTION: Biological performance as a result of executing a behavioral command.

BARYONS: In the standard model, all particles that are composed of 2 or more quarks. A quark has a baryon number of 1/3. Antiquarks have a baryon number of −1/3. Baryon number is conserved in all interactions.

BEHAVIOR: The biological performance of a set of perception, decision and action (PDA) cycles. The components of behavior are the three related processes of perception, decision and action.

BEHAVIORAL COMMAND: An instruction designed to initiate the operational performance of a cell or a group of cells.

BIG BANG: The experimental finding that the universe began to expand from a very small region of space and time, giving birth to galaxies which are all receding from each other. This recessional motion is now known to be an accelerated motion.

BIOENERGY: Energy which is living.

BODY SPIRIT: The electromagnetic field produced by charge and motion of charge associated with the molecular structure of bioenergy. The body spirit does not include the controlling induced electromagnetic fields (the soul).

BOSON: A classification of elementary particle whose spin angular momentum can only take on integral values of Planck's constant. Bosons are the force carrying particles that make up force fields. See also force fields.

CELL: A fundamental building block of bioenergy. The most fundamental building block of the cell nucleus (the central controlling portion of the cell) is the DNA molecule.

CELL NUCLEUS: The central controlling portion of one biological cell. All cells belonging to any multicelled organism contains identical nuclei except the sex cells which contain identical half nuclei.

CENTRAL FISSURE: A fissure which separates the front portion from the back portion of the human brain. The front portion of the human brain controls the muscles for motor control. The back portion of the human brain receives the incoming signals which comprise perception.

CHARGE: Electric charge can either be positive or negative. Macroscopic charge is quantized as multiples of the charge on the electron. Protons and positrons carry a single positive charge, while electrons carry a single negative charge. Microscopic electric charge carried by quarks is quantized as 1/3 or 2/3 electron charge.

CHRISTIAN TRINITY: The Christian philosophy that God exists as three distinct quantities, the

Father (as Jesus Christ's Living Personal God), Son (as the Living Soul of Jesus Christ) and connective part of the Holy Spirit between Jesus and The Father. Together, the energy of The Father, Son and connective energy (part of the Holy Spirit) between the Father and Son, make up the total invariable energy in the universe (God). The Personal Living God of each human has exactly the same Mind and Soul as The Father. See also Living Personal God. See also Holy Spirit. See also Soul of God.

CLONING: Reproduction resulting in a material duplicate of the parent.

COLOR: Quarks can either be red, green or blue which is responsible for the strong nuclear force (forces between quarks) much like charge being responsible for the electric coulomb force. It can be shown that a linear combination of these three colors can form the normal quantized colorless charge of the electron, positron or atomic nucleus.

CONSCIOUSNESS: The real time (or present time) cyclic performance of perception, decision and resulting action.

CORPUS COLLOSSUM: A fissure which separates the left hemisphere from the right hemisphere of the human brain. The left hemisphere is connected to the right portion of the human body. The right hemisphere is connected to the left portion of the human body.

DECISION: The process of formulating a behavioral command.

DEMAND: The potential for the consumption of supply.

DIRECTIVE SURVIVAL: Behavior which enhances the overall conservation and preservation of bioenergy. Directive survival for the individual means the satisfaction of the generalized drive. Directive survival for the group means survival of the species.

DNA MOLECULE: The fundamental living molecule which rerprents the genetic information usually found inside the cell nucleus (deoxyribose nucleic acid).

DYNAMIC: Dependent on time.

ENERGY: One of the two constituents of the universe (as opposed to space). Mass and charge are the two fundamental independent energy descriptors.

ENERGY CONSERVATION: The experimental finding that energy cannot be destroyed or created.

ENERGY DESCRIPTOR: An experimentally defined unit of mass or charge.

EVENT: A representation of the position(s) of a dynamic energy system(s) by a point(s) in four dimensional space. The specification of the behavior, existence or transformation of an energy system.

FATHER: Every human being's Personal Living God, as well as Jesus Christ's Personal Living God. See also Christian Trinity. See also Personal Living God.

FERMION: A classification of elementary particle whose spin angular momentum can only take on half integral values of Planck's constant.

FIELD: A force field.

FIELD ENERGY: The energy associated with a force field.

FIELD OF CHARGE: An electric or magnetic field.

FIELD OF MASS: A gravitational or nuclear field.

FIRST CAUSE: A perception of God as being the primal force behind the creation of the universe.

FISSION: Breaking apart of a heavy atomic nucleus into lighter atomic nuclei and accompanied by other energetic particles.

FORCE FIELD: Energy composed of bosons capable of influencing the motion of mass and charge. The exchange of bosons is the root cause of forces.

FREE WILL: The fact that the behavior of bioenergy cannot be exactly predicted or known (measured) and is a consequence of the Heisenberg uncertainty principle.

FUSION: Combining light atomic nuclei into a heavier atomic nucleus and accompanied by other energetic particles.

GENERALIZED DRIVE: The primal biological desire/need for gas, water, food and reproduction.

GLUON: A boson responsible for the strong nuclear force which binds the quarks together in protons and neutrons.

GOD: All the energy in the universe. This includes all material energy, as well as all field energy. It includes all living energy as well as non-living energy. Experimentally, it is found that energy cannot be created or destroyed, thus God is the invariable universal total energy.

GRAVITON: The boson responsible for the gravitational force between any two masses.

HARDWARE: Composed of matter.

HEISENBERG UNCERTAINTY PRINCIPLE: The momentum and position of any dynamic energy system cannot be known simultaneously. The energy and when the system had that energy cannot be known simultaneously.

HELICITY: The component of a particles spin angular momentum in the direction of the particle's velocity vector.

HOLY SPIRIT: The connective, induced electro-magnetic field energy between God's Soul and all other living souls. The Holy Spirit connects your soul with the Soul of God. The Holy Spirit connects Jesus Christ's Soul with the Father defined to be Jesus Christ's Personal Living God. See also Soul of God. See also Personal Living God.

IMMORAL BEHAVIOR: Behavior which is not in accord with directive survival.

INTERACTION: Influence on the motion of an energy system via a force field.

KINETIC: The state of relative motion (dynamic) (as in kinetic energy).

KNOWLEDGE: The ability to describe or predict measurable events.

LEPTONS: In the standard model, all matter is composed of 3 families of 4 particles. Each family (F) contains 2 quarks and 2 leptons. The families are:

(F1) up and down quarks and associated leptons (electron, electron antineutrino),

(F2) charmed and strange quarks and associated leptons (muon, muon antineutrino),

(F3) top and bottom quarks and associated leptons (tauon, tauon antineutrino).

Lepton number is conserved in all interactions. See also standard model.

LIVING(NESS): A dynamic energy system arranged in such a way that it can behave. It must contain some induced controlling electromagnetic field energy and it must contain some static electromagnetic connective energy to be alive. The dynamic microscopic DNA molecule is the most fundamental genetic hardware component of life.

The dynamic cell is the most fundamental macroscopic unit of life.

LONG TERM MEMORY: The set of all neurons which contain the record of all past experiences.

MATTER: Stored energy which takes up space and has mass.

MEASUREMENT: A comparison against the standard.

MEMORY: The set of neurons responsible for the stored experiences (biological). A device which can store information (computer).

MORAL BEHAVIOR: Behavior in accord with directive survival.

MOTOR CONTROL: The front portion of the human brain is responsible for sending electrical signals to the muscles of the body that control motion.

NEURON: A specialized cell capable of generating or conducting electrical current. The fundamental unit of the nervous system.

NEUTRON: Consists of 2 down quarks and 1 up quark. Each down contains $-1/3$ electric charge. The up quark contains $+2/3$ electric charge. Its net charge is 0.

NUCLEUS: The central portion of an atom or biological cell. The nucleus of a biological cell contains the genetic DNA molecules. The nucleus of an atom generally contain protons and neutrons. Each atom is distinguished from another by its number of protons.

PDA CYCLE: The repetitious process of perception, decision and action. If the cycle occurs in present time, then this process forms consciousness. If perception involves vividly imagined stimulus (internal stimulus), the process forms unconscious behavior or simulated behavior.

PERCEPTION: External and internal detection is by direct physical contact. Simulated detection is by vividly imagined physical contact. See also simulated behavior.

PERSONAL LIVING GOD: The total energy of the universe less the energy representing any human's soul including its connective energy part of the Holy Spirit represents that human's personal living God. The energy of any human's soul plus the energy of its personal living God plus the connective energy of the Holy Spirit represents God, the invariable total. Thus, any human's soul interacts with its own personal living God via the Holy Spirit. Each and every personal living God is the same as the Father (Jesus Christ's Personal Living God) since every personal living God contains the single Soul and Mind of God. See also Christian Trinity. See also Holy Spirit. See also Soul of God.

PHOTON: A particle of light having no rest mass. This boson is responsible for the forces between charges. The frequency of the induced electromagnetic wave corresponds to a photon in flight and is proportional to its energy. The constant of proportionality is Planck's constant.

PIONS: The 3 bosons responsible for the forces holding neutrons and protons together in all atomic nuclei. These are the pi plus, pi minus and pi zero. See also standard model.

PLANCK'S CONSTANT: When divided by 2π is symbolized by $\hbar$. It is the smallest unit of spin angular momentum. Fermions have half integral multiples of $\hbar$. Bosons have integral values of $\hbar$.

POTENTIAL: The state of relative position (static) (as in potential energy).

PROTON: Consists of 2 up quarks and 1 down quark. The down contains $-1/3$ electric charge. Both up quarks contains $+2/3$ electric charge. Its net charge is $+1$.

QUARKS: In the standard model, there are 3 families of 2 quarks. These are the up and down quarks, the charmed and strange quarks and the top and bottom quarks. See also standard model.

REALITY: The set of all measurable events.

SELF IMAGE MEMORY: The portion of long term memory which is self consistent. It is also known as the subconscious mind. It defines what is possible behavior for the composite self.

SHORT TERM MEMORY: The set of neurons containing all the information which is currently being processed.

SIMULATED BEHAVIOR: The process of vividly imagined behavior. The process of picturing oneself performing desired behavior and thereby (re)programming self image memory.

SON: The Living Soul of Jesus Christ. See also Christian Trinity. See also Personal Living God. See also Universal Spirit, Soul and Soul of God.

SOUL: The induced (living) electromagnetic field energy which controls the electromagnetic body spirit of any living entity. See also body spirit.

SOUL OF GOD: The mind or living part of the induced electromagnetic field energy that existed both before and after the big bang. The pre-bang Universal Spirit was the Soul of God (a living mind) plus the connective induced electromagnetic energy of the Holy Spirit. See also Holy Spirit. See also Universal Spirit. See also Personal God.

SPACE: One of the two constituents of the universe (as opposed to energy). Length and time are the two fundamental independent space or position descriptors.

SPACE DESCRIPTOR: The fundamental measurable space quantities of position and time. An experimentally defined unit of length or time.

STANDARD: A unit of measure which has been agreed upon.

STANDARD MODEL: All matter is composed of three families of four fermion particles each. The first two members are quarks and the last two members are leptons. These matter families (M) are:

M1 (up quark, down quark, electron, electron anti–neutrino)

M2 (charmed quark, strange quark, muon, muon anti–neutrino)

M3 (top quark, bottom quark, tauon, tauon anti–neutrino)

All field energy is composed of three families of three boson particles each.

The first family F1 includes the gluons responsible for the force that holds the quarks in both neutrons (2 down quarks & 1 up quark) and protons (2 up quarks & 1 down quark) together. The photon is responsible for forces between charges and the graviton is responsible for forces between masses.

The second family F2 includes the 3 pions responsible for holding neutrons and protons of atomic nuclei together.

The third family F3 includes the 3 weakons responsible for forces causing radioactive decay. These three field families (F) are:

F1 (gluon, photon, graviton),

F2 (pi plus, pi minus, pi zero),

F3 (omega plus, omega minus, zeta zero)

All matter particles also have a partner which is called its anti–particle. Anti–particles have the same mass as the particles but the sign of their charge and helicity is reversed. See Basic Elementary Particle Section.

SUPPLY: All consumable goods and services in the market.

SURVIVAL: Conservation and preservation of bioenergy. See Directive Survival.

SURVIVAL VALUE: Human knowledge which enhances human life or increases its ability to survive is termed survival value. More survival value means more effective control of humanity over its environment.

TRUTH: A subset of reality.

UNCONSCIOUSNESS: The performance of internal (vividly imagined) behavior (perception, decision and action) resulting in the (re)programming of long term memory. See also simulated behavior. See also consciousness.

UNIVERSAL DESCRIPTOR: The fundamental measurable energy or space quantities (mass, charge, lengths and time).

UNIVERSAL SPIRIT: All the force field energy in the universe including the controlling, directing field (Soul of God). The Soul of God and Holy Spirit were the only existing induced electromagnetic field energy in the universe before the big bang. See also Soul of God. See also Holy Spirit.

VECTOR: A physical quantity having both magnitude and direction.

WEAKONS: The three bosons responsible for the electroweak force. In the standard model of elementary particles, these are the omega plus, the omega minus and the zeta zero. The electroweak force is responsible for the decay of radioactive elements. See also standard model.

Fundamental Physical Laws

Preliminary Definitions

I. **Bold** mathematical single letters refer to vectors.

II. The symbol, $i = (-1)^{1/2}$ always occurs as the fourth (time component) of all Einsteinian four dimensional vectors.

III. The symbol, c is the speed of light in vacuum.

1. **Position of an energy system:** Referring to Figure 2, a normal Cartesian coordinate system shows the (x,y,z) position of the system S at time t.

Newtonian: $\mathbf{r}_N = (x,y,z)$

Einsteinian: $\mathbf{r}_E = (x,y,z,ict)$

2. **Velocity of an energy system:** At time t_2, the position of the system was at position 2, (r_2). Initially at time t_1, the position of the system was at position 1, (r_1). The average velocity of the system is the distance traversed by the system in moving from position 1 to position 2 ($r_2 - r_1$) divided by the time it took for the system to move between the two positions ($t_2 - t_1$). The direction of the velocity is from position 1 to position 2. The instantaneous velocity $\mathbf{v}$ is realized by letting t_2 approach t_1. Mathematically, the instantaneous velocity of a system is a vector quantity.

$\mathbf{v} = \lim$ as $t_2 \rightarrow t_1$ of $[(\mathbf{r_2} - \mathbf{r_1})/(t_2 - t_1)]$ or

$\mathbf{v} = d\mathbf{r}/dt$

Newtonian: $\mathbf{v}_N = (v_x, v_y, v_z)$

Einsteinian: $\mathbf{v}_E = (v_x, v_y, v_z, ic)$

3. **Acceleration of an energy system:** At time t_2, the velocity of the system was (v_2). Initially at time

t_1, the velocity of the system was (v_1). The average acceleration of the system is the change in the velocity of the system in going from v_1 to v_2, ($v_2 - v_1$) divided by the time it took for the system to go from v_1 to v_2, ($t_2 - t_1$). The direction of the acceleration is from v_1 to v_2. The instantaneous acceleration **a** is realized by letting t_2 approach t_1. Mathematically, the acceleration of a system is a vector quantity.

$\mathbf{a} = \lim$ as $t_2 \to t_1$ of $[(\mathbf{v_2} - \mathbf{v_1})/(t_2 - t_1)]$ or

$\mathbf{a} = d\mathbf{v}/dt$

Newtonian: $\mathbf{a}_N = (a_x, a_y, a_z)$

Einsteinian: $\mathbf{a}_E = (a_x, a_y, a_z, 0)$

4. **Momentum of an energy system:** The product of the system's mass and its velocity **v** is called its momentum and denoted by **p**. It is a vector quantity having direction **v**.

$\mathbf{p} = m\mathbf{v}$

Newtonian: $\mathbf{p}_N = (p_x, p_y, p_z)$

Einsteinian: $\mathbf{p}_E = (p_x, p_y, p_z, iE/c)$

where E is the total energy $= mc^2$.

5. **Force on an energy system:** The instantaneous rate of change of a system's momentum with respect to time. Its definition is similar to the definition of velocity. At time t_2 it has momentum 2. At time t_1, it had momentum 1. The average force is the difference in momentum (momentum 2 – momentum 1) divided by the time difference $(t_2 - t_1)$. The instantaneous rate is realized when t_2 approaches t_1.

$\mathbf{F} = \lim$ as $t_2 \rightarrow t_1$ of $[((m\mathbf{v})_2 - (m\mathbf{v})_1)/(t_2 - t_1)]$ or

$\mathbf{F} = d(m\mathbf{v})/dt$

Newtonian: $\mathbf{F}_N = m_0 d\mathbf{v}/dt = m_0 \mathbf{a}$

where m_0 is the rest mass and **a** is the acceleration.

Einsteinian: $\mathbf{F}_E = d(m\mathbf{v})/dt = md\mathbf{v}/dt + \mathbf{v}dm/dt$

$$= m_0\mathbf{a}(1- v^2/c^2)^{-3/2} = (c^2/v)dm/dt$$

where **v** is m's velocity and **a** is m's acceleration.

6. **Mass density:** Mass m, per unit volume V.

Average mass density = ρ_{mavg} = m/V

Instantaneous mass density = ρ_m = dm/dV

7. **Pressure on a surface**: The applied force F, per unit surface area, A.

Average pressure = P_{avg} = F/A

Instantaneous pressure = P = dF/dA

8. **Angular momentum of an energy system:** Let the vector from the origin to the position of the

system be called the position vector (**r**). The angular momentum of the system (**L**) is then the ordinary vector cross product (**x**), of the position vector with the system's momentum vector (m**v**).

$$L = r \ x \ mv$$

9. **Charge density:** Amount of charge q, per unit volume V.

Average charge density = ρ_{qavg} = q/V

Instantaneous charge density = ρ_q = dq/dV

10. **Electrical current:** The instantaneous change of charge q with respect to time.

i = lim as $t_2 \rightarrow t_1$ of [$(q_2 - q_1)/(t_2 - t_1)$]   or

i = dq/dt

11. **Electrical current density:** The electrical current i per unit cross sectional area A of

conductor. The unit vector **u** has a direction of the current i along the conductor perpendicular to the cross sectional area.

$$\mathbf{J}_i = \mathbf{u}i/A \text{ where } i = dq/dt$$

In a conductor with conductivity σ_c, the current is in the direction of the electric field **E** and the current density is the product of the conductivity and electric field.

$$\mathbf{J}_i = \sigma_c\mathbf{E}$$

12. **Mole**: The mass of Avogadro's number of identical molecules or Avogadro's number of identical atoms expressed in grams. One mole of molecules is the molecular weight of the molecule expressed in grams. One mole of atoms is the atomic weight of the atom expressed in grams.

Mechanical Laws

1. Newton's Laws of Motion:

1.1   A body will remain at rest or in motion at a constant velocity unless acted on by an unbalanced external force.

1.2   The force on a body is proportional to its acceleration and the constant of proportionality is the rest mass (when the body is at rest), m_0 of the body.

$\mathbf{F} = m_0\mathbf{a}$

(Newton was unaware that mass is a function of its velocity.)

1.3   The force of one body on a second body is equal and opposite to the force of the second body on the first body or for every action, there is an equal and opposite reaction.

$$\mathbf{F}_{12} = -\mathbf{F}_{21}$$

1.4   Newton's Universal Law of Gravitation says that any two energy systems having mass attract each other with a force ($\mathbf{F}$) proportional to the product of their masses m_1 and m_2 and inversely proportional to the square of the distance (r) between their mass centers. The force is in a direction between the centers of m_1 and m_2, causing them to attract one another and is denoted by the unit vector $\mathbf{r}_u$. G is the constant of proportionality known as Newton's Gravitational Constant. This force is

$$\mathbf{F} = \mathbf{r}_u Gm_1m_2/r^2$$

deriving the gravitational potential energy, V between m_1 and m_2 as

$$V = -\mathbf{G}m_1m_2/r$$

Newtonian: Mass is the cause of the gravitational field.

Einsteinian: Mass energy and momentum warp four dimensional spacetime into a gravitational field.

2. Quantum Mechanical Laws:

2.1 An energy system may be described by a wave function. The total energy operator $\hat{H}$ (known as the Hamiltonian) operating on the wave function (Ψ) yields the total energy eigenvalue (E) of the system represented by the wave function. Energy eigenvalues (E) are the allowable energy states that the system may assume. Similarly, other operators operating on the wave function yield other information (such as the spin, momentum, angular momentum, etc.) about the system.

$$\hat{H}\Psi = E\Psi$$

2.2 The square of the wave function $\Psi^*\Psi$, multiplied by a infinitesimal volume d^3r is equal to the infinitesimal probability dP, that a system specified by Ψ, is located within that volume.

$$dP = \Psi^*\Psi d^3r$$

2.3   The probability that an energy system represented by the wave function Ψ, is somewhere in all space is unity, which is the basis for a normalized wave function.

$$P = \int dP = \int \Psi^*\Psi d^3r = 1$$

3. The Heisenberg Uncertainty Principle:

3.1   In an ideal experiment, the product of the standard deviation in the measurement of a system's momentum, Δp and the standard deviation in the measurement of its position, Δr must be greater than a non-zero constant. This constant is Planck's constant divided by four pi ($h/(4\pi) = \hbar/2$) where $\hbar$ is Planck's constant divided by 2π.

$$\Delta p \Delta r \geq \hbar/2$$

This means that an energy system's position and momentum cannot be known simultaneously.

3.2   Another expression of the Heisenberg uncertainty principle is:

$$\Delta E \Delta t \geq \hbar/2$$

where ΔE is the standard deviation in the measurement of a system's energy and Δt is the standard deviation of the measured times that it had that energy. This means that a system's energy and when it had that energy cannot be known simultaneously.

4. **The energy of a photon (E_γ):** an electromagnetic wave's energy is either the product of its frequency ω, and Planck's constant $\hbar$, or its mass m_γ, and the speed of light, c squared.

$$E_\gamma = \hbar\omega = m_\gamma c^2$$

5. **De Broglie's relationship:** which expresses that the wavelength of a particle λ is inversely proportional to its momentum. The constant of proportionality is Planck's constant, h.

$\lambda = h/mv$

and is sometimes written as

$\lambdabar = \hbar/mv$

where $\lambdabar = \lambda/(2\pi)$ and $\hbar = h/(2\pi)$

6. **Einstein's Laws of Special Relativity:** The first four relativistic laws are derived by assuming that the velocity of light c, is independent of the velocity of the source of light as well as the velocity of the observer.

6.1   A system's mass m increases if it is moving with a velocity v compared to the velocity of light c, in vacuum. Initially when the system had a velocity of zero, its rest mass is m_0.

$$m = m_0 \left(1 - (v/c)^2\right)^{-1/2}$$

6.2   A system's length ℓ decreases if it is moving with a velocity v compared to the velocity of light c, in vacuum. Initially when the system had a velocity of zero, its rest length is ℓ_0. ℓ is in the direction of the velocity.

$$\ell = \ell_0 \left(1 - (v/c)^2\right)^{1/2}$$

6.3   A system's clock time length, t slows (stretches) if it is moving with a velocity v compared to the velocity of light c, in vacuum. Initially when the system was at rest (had a velocity of zero), it had a clock time length of t_0.

$$t = t_0 \left(1 - (v/c)^2\right)^{-1/2}$$

6.4   The total mechanical energy E of a system containing mass is the product of its mass m and the square of the velocity of light c.

$$E = mc^2$$

where $m = m_0(1 - (v/c)^2)^{-1/2}$ is dependent on its velocity v. m_0 (rest mass) is its mass when $v = 0$.

6.5 The relativistic kinetic energy T, of a system in motion is the difference (between its mass in motion less its rest mass) times the velocity of light c, squared.

Einsteinian: $T_E = (m - m_0)c^2$

where $m = m_0(1 - (v/c)^2)^{-1/2}$ is dependent on its velocity, v. For small velocities compared to the velocity of light, the Einsteinian kinetic energy reduces to the Newtonian kinetic energy as a first order approximation. For $v \ll c$, $(m - m_0)c^2 \cong (\frac{1}{2})m_0v^2$ producing the classic kinetic energy.

Newtonian: $T_N = (\frac{1}{2})m_0v^2$

7. **Laws of Thermodynamics**

7.1   The first law of thermodynamics says that within a closed (isolated) system an amount of heat added to the system dQ results in an increase in its internal energy dU and an amount of work done, dW. Usually, dU results in an increase in internal temperature while dW results in a change in volume dV against a constant pressure p. This also means that energy is conserved for a closed system.

$dQ = dU + dW$ where $dW = pdV$

7.2 The second law of thermodynamics says that a change in the entropy dS of a system undergoing a reversible process is defined to be the amount of heat added dQ divided by its temperature T. If the process is irreversible, then the entropy is always greater than the amount of heat added divided by its temperature.

$dS \geq dQ/T$

where the equality implies reversibility and the greater than symbol (>) implies irreversibility.

7.3 The perfect gas law says that the gas pressure p multiplied by the volume of gas V is proportional to the number of moles n of gas multiplied by the absolute temperature T of the gas. The constant of proportionality R is known as the universal gas constant.

$pV = nRT$

7.4 The fundamental law of heat conduction says that the rate of heat flow dQ/dt across a infinitely thin slab dx of material perpendicular to the surface of the slab is proportional to the surface area A of the slab and the instantaneous absolute temperature change per unit thickness dT/dx of the material. The constant of proportionality K_T is known as the thermal conductivity of the material. The minus sign means that heat flow is in a direction of decreasing temperature.

$$dQ/dt = -K_T A \, dT/dx$$

7.5   The internal energy U of an ideal gas containing N molecules is proportional to the product of N and the absolute temperature T. The constant of proportionality is 3k/2 where k is Boltzmann's constant.

$$U = (3/2)NkT$$

7.6   In an idealized heated solid called a cavity radiator, the energy radiated from the cavity interior per unit area (called total cavity radiancy, R_C) is proportional to the fourth power of the absolute temperature T. The constant of proportionality σ is called the Stefan-Boltzmann constant.

$$R_C = \sigma T^4$$

8. Temperatures and Conversions

C^0 is the symbol for degrees Celsius, F^0 is the symbol for degrees Fahrenheit and K^0 means degrees Kelvin (Absolute).

8.1 Water freezes at 0 C^0 at standard atmospheric pressure.

8.2 Water boils at 100 C^0 at standard atmospheric pressure.

8.3 The triple point of water (existing simultaneously as a gas, liquid and solid) occurs at a temperature of 273.16 K^0 and atmospheric pressure of 611.73 Pascals (Newtons per square meter).

8.4 $C^0/100 = (F^0 - 32)/180$

8.5 $K^0 = C^0 + 273.16$

Electromechanical Laws

1. Maxwell's Equations:

1.1   The source of the electric field ($\mathbf{E}$) is charge density ρ_q. $\nabla = (\partial/\partial x, \partial/\partial y, \partial/\partial z)$ is the normal vector operator, ($\bullet$) is the normal vector scalar product and ε_0 is a constant called the permittivity of free space. This law is also known as Gauss's law for electricity. The differential form is

$$\nabla \bullet \mathbf{E} = \rho_q/\varepsilon_0$$

The integral form is

$$\varepsilon_0 \oiint \mathbf{E} \bullet \mathbf{n} dS = q$$

where $\oiint$ means integration over the closed surface S, $\mathbf{n}$ is a unit vector normal to S enclosing the charge q.

1.1.1 Maxwell's first equation and may be used to derive Coulomb's law which states that the force between two charges is proportional to the product of the two charges and inversely proportional to the square of the distance between their charge centers. The force is in a direction on a line drawn between the two charges q_1 and q_2 denoted by the unit vector $\mathbf{r}_u$. $K_C = 1/(4\pi\varepsilon_0)$ will be called Coulomb's constant.

$$\mathbf{F} = \mathbf{r}_u K_C q_1 q_2 / r^2$$

giving rise to the electrical potential energy, V between q_1 and q_2

$$V = K_C q_1 q_2 / r$$

If the charges are both positive or both negative, the force is repulsive (like charges repel one another), otherwise the force is attractive (unlike charges attract one another).

1.2 The source of the magnetic field $\mathbf{B}$ is zero. This is Maxwell's second equation. This also means that magnetic fields always exist in closed loops and magnetic monopoles do not exist. This law is also known as Gauss's law for magnetism. The differential form is

$$\nabla \bullet \mathbf{B} = 0$$

The integral form is

$$\oiint \mathbf{B} \bullet \mathbf{n} dS = 0$$

where $\oiint$ means integration over any closed surface S, $\mathbf{n}$ is a unit vector perpendicular to the surface, S.

1.3 Ampere's law is also known as Maxwell's third equation. Current density $\mathbf{J}_i$ and/or dynamic electric fields $\partial \mathbf{E}/\partial t$ give rise to circulating magnetic fields $(\mathbf{B})$. μ_0 is known as the permeability constant of free space. The differential form is

$$\nabla \times \mathbf{B} = \mu_0 \mathbf{J}_i + \mu_0 \varepsilon_0 \partial \mathbf{E}/\partial t$$

where $\nabla = (\partial/\partial x, \partial/\partial y, \partial/\partial z)$ is the Del vector operator and $\mathbf{x}$ is the vector cross product. The integral form is

$$(1/\mu_0)\oint \mathbf{B} \bullet \mathbf{ds} = i$$

where $\oint$ means integration over a closed line s, circulating around the electrical current i. $\mathbf{ds}$ is a vector line element on the line, s and perpendicular to the direction of the electrical current i.

1.4   Faraday's law is also known as Maxwell's fourth equation. Dynamic magnetic fields ($\partial \mathbf{B}/\partial t$) give rise to circulating electric fields ($\mathbf{E}$). The differential form is

$$\nabla \mathbf{x} \mathbf{E} = -\partial \mathbf{B}/\partial t$$

where $\nabla = (\partial/\partial x, \partial/\partial y, \partial/\partial z)$ is the normal Del vector operator and $\mathbf{x}$ is the vector cross product.

The integral form is

$$\oint \mathbf{E} \bullet \mathbf{ds} = -\iint (\partial \mathbf{B}/\partial t) \bullet \mathbf{n} dS = -\partial \Phi/\partial t$$

where $\Phi = \iint \mathbf{B} \bullet \mathbf{n} dS$ is called the magnetic flux in which $\mathbf{B}$ penetrates the surface area S. $\mathbf{n}$ is a unit vector perpendicular to the surface area S.

2. The Lorentz Force:

The force $\mathbf{F}$ on a charge q moving with velocity $\mathbf{v}$ by an external electric field $\mathbf{E}$ and by an external magnetic field $\mathbf{B}$ and $\mathbf{x}$ is the normal vector cross product.

$$\mathbf{F} = q\mathbf{E} + q\mathbf{v} \ \mathbf{x} \ \mathbf{B}$$

3. Electromagnetic Wave Equations:

When there is no charges or currents, as in the vacuum of matter free space, Maxwell's equations yield a wave equation that is satisfied by both the electric field $\mathbf{E}$ as well as the magnetic field $\mathbf{B}$.

These equations yields the precise description of induced electromagnetic fields.

3.1 $\nabla^2 \mathbf{E} - \partial^2 \mathbf{E}/(c^2 \partial t^2) = 0$ and

3.2 $\nabla^2 \mathbf{B} - \partial^2 \mathbf{B}/(c^2 \partial t^2) = 0$

where $\nabla^2 = \nabla \bullet \nabla = \partial^2/\partial x^2 + \partial^2/\partial y^2 + \partial^2/\partial z^2$, t is the time and c is the speed of light in vacuum. Note that if one utilizes the Dalembertian operator, $\square^2 = \partial^2/\partial x^2 + \partial^2/\partial y^2 + \partial^2/\partial z^2 - \partial^2/(c^2 \partial t^2)$ the electromagnetic wave equations 3.1 and 3.2 simplify to

3.1.1 $\square^2 \mathbf{E} = 0$ and

3.2.1 $\square^2 \mathbf{B} = 0$

Conservation Laws

1. Conservation of energy: A system's total energy, E_T is the same both before (B) and after (A) any energy transformation.

$(E_T)_B = (E_T)_A$

2. Conservation of momentum: A system's total momentum, p_T is the same both before and after any energy transformation.

$(p_T)_B = (p_T)_A$

3. Conservation of angular momentum: A system's total angular momentum, L_T is the same both before and after any energy transformation.

$(L_T)_B = (L_T)_A$

4. Conservation of charge: A system's total charge, Q_T is the same both before and after any energy transformation.

$$(Q_T)_B = (Q_T)_A$$

5. Conservation of baryon number: A system's baryon number, N_B is the same both before and after any energy transformation. Baryons are composed of quarks. Quarks have baryon number $+1/3$. Antiquarks have baryon number $-1/3$.

$$(N_B)_B = (N_B)_A$$

6. Conservation of lepton number: A system's lepton number, N_L is the same both before and after any energy transformation.

$$(N_L)_B = (N_L)_A$$

7. For any energy system, another related energy system predicted by the simultaneous operations of time reversal, charge conjugation (signs of all charges involved are reversed) and space reversal

(mirror image or parity) is also possible. This is called CPT for short. Below, E_T is the total energy of a system and BCPT means before the CPT operation and ACPT means after the CPT operation.

$$(E_T)_{BCPT} = (E_T)_{ACPT}$$

Basic Units

position: (measured with a ruler)

meter = m

mass: (measured with a balance scale)

kilogram = kg

time: (measured with a clock)

second = s

charge: (measured with a voltmeter)

coulomb = coul

Equivalent Units

Force: Newton = nt = $kg–m/s^2$

Pressure: Pascal = nt/m^2

Energy: joule = nt–m

Inductance: henry = $joule–m–s^2/coul^2$

Capacitance: farad = $coul^2/joule$

Basic Physical Constants

Name	Symbol	Value
Speed of light	c	3.00×10^8 m/s
Gravitational Constant	G	6.67×10^{-11} nt–m^2/kg^2
Avogadro's number	N_0	6.023×10^{23} /mole
Universal Gas Constant	R	8.32 joules/(mole–K^0)
(Planck's constant)/2π	$\hbar$	1.055×10^{-34} joule–s
Planck length	$L_P = (\hbar G/c^3)^{1/2}$	1.616×10^{-35} m
Planck time	$T_P = (\hbar G/c^5)^{1/2}$	5.391×10^{-44} s
Planck mass	$M_P = (\hbar c/G)^{1/2}$	2.177×10^{-8} kg
Boltzmann's constant	k	1.38×10^{-23} joules/(molecule–K^0)
Stefan-Boltzman constant	σ	5.67×10^{-8} joules/m^2/(K^0)^4
Permeability constant	μ_0	1.26×10^{-6} henry/m
Permittivity constant	ε_0	8.85×10^{-12} farad/m

Name	Symbol	Value
Electron charge	q_e	-1.6022×10^{-19} coul
Electron rest mass	m_e	9.11×10^{-31} kg
Proton rest mass	m_p	1.67239×10^{-27} kg
Neutron rest mass	m_N	1.6747×10^{-27} kg
Coulombs constant	$1/(4\pi\varepsilon_0)$	8.99×10^9 nt–m^2/coul^2

Basic Elementary Particles

Preliminary Particle Descriptors

1. Family Names – Particles belong to functional families having a set number of family members. For example, the gluon family has eight members and they function to provide the strong nuclear force that hold quarks together. Individual particles have both a historical name and a symbol. For example, an electron has the symbol e^-.

2. Color – Quarks can either be red, green or blue (r,g,b). Anti-quarks can either be –red, –green or –blue (–r, –g, –b). This is similar to charge coming in two types, the minus (–) and the plus (+) type.

3. Charge – measured in units of positive electronic charge or the charge on a positron (anti–electron). The charge magnitude of a negative electron (e^-) or a positive positron (e^+) are equal. An anti–particle has the opposite charge as the particle.

4. Spin – Axial angular momentum measured in units of Planck's constant divided by 2π and denoted by $\hbar$. Quantum Spin is specified as positive, but it is understood that quantum mechanically, it can either be positive (parallel) or negative (anti–parallel) to any given direction. Fermions (matter particles) have half integral values of $\hbar$. Bosons (force field particles) have integral values of $\hbar$.

5. Helicity – Helicity is also given in terms of $\hbar$ and may be thought of as the component of the particle's spin in the direction of the particle's velocity vector. The helicity of particles moving at the velocity of light is different than the helicity of particles that do not. Particles moving at the velocity of light, c such as photons, must have zero rest mass and there is no coordinate system for which its velocity is zero. Thus, the component of a photon's spin ($\hbar$) along its velocity vector is the same as its spin orientation, either $+\hbar$ or $-\hbar$ since it cannot be observed at rest. Thus, a photon has an intrinsic helicity the same as

its intrinsic spin. On the other hand, particles with non-zero rest mass have non–intrinsic helicity dependent on the observer since their spin can be observed when they are at rest and their spin components in the direction of motion must have a quantum difference of $+\hbar$. For example, the weakons, responsible for the electroweak forces, with non-zero rest masses and spin of $\hbar$ have helicity of either $-\hbar$, 0, or $+\hbar$. A particle and its anti–particle have opposite helicity.

6. Rest Mass – Measured in either Proton rest masses (Mp) or millions of electron volts (Mev). An electron volt (1.602×10^{-19} joules) is the kinetic energy an electron gains by being propelled a distance of one meter by an electrical field of strength, one volt per meter. The equivalent energy of a proton at rest is 938 Mev. The reason rest mass can be measured in terms of energy is because of Einstein's famous equation $E_0 = m_0c^2$ which relates rest mass, m_0 to rest mass energy, E_0 by a constant, being the square of the speed of light, c^2.

7. Field Energy – Force fields are caused by corresponding field particles having integral values of $\hbar$ (called bosons). Matter particles having half integral values of $\hbar$ (called fermions) are influenced by force fields caused by their interaction with the corresponding boson. The four force fields are strong nuclear (gluons), electroweak (weakons), electromagnetic (photons) and gravitational (gravitons).

Anti–Particle Properties

All particles have an anti–particle. The anti–particle has the opposite charge of the particle. The anti–particle has the opposite helicity of the particle. The anti–particle of a non–zero rest mass particle having zero charge, and having a spin of one $\hbar$ and zero helicity is the particle itself. The antiparticle has the same mass as the particle. A particle and its anti–particle (that is not itself) annihilate one another upon contact in a burst of other energetic particles.

Matter Energy Particles

All material energy is composed of fundamental matter particles experimentally observed to exist as three energy families (UP, CHARMED, TOP) of four fermions each, in its simplest representation. Two of the fermions are light and are called leptons and two of the fermions are heavy and are called quarks. One of the leptons carries a negative electronic charge, the other has no charge.

Origin of the UP Family

The nuclei of atoms are composed of neutrons and protons. A neutron consists of two (red and blue) down quarks, $(d_R^{-1/3},\ d_B^{-1/3},\ u_G^{2/3})$ and one (green) up quark, . A proton consists of two (red and blue) up quarks, and one (green) down quark, $(u_R^{2/3},\ u_B^{2/3},\ d_G^{-1/3})$. Any other cyclic permutation of red, green or blue colored quarks in neutrons or protons is possible. The proton is stable. An isolated neutron, n is unstable and will decay into a proton, p electron, e^- and an electron anti–neutrino, $\acute{\upsilon}_e$. The net effect is that one of the down quarks of the neutron will change into an electron, anti–neutrino and an up quark. This effectively transformed the internal structure of a neutron ($d_R^{-1/3}$, $d_B^{-1/3}$, $u_G^{2/3}$) into that of a proton ($u_R^{2/3}$, $u_B^{2/3}$, $d_G^{-1/3}$). The up quark has a charge of 2/3 e^+ while the down quark has a charge of $-1/3$ e^+. Thus a proton has a net charge of e^+ while the neutron has a net charge of 0. The UP family making up neutrons and protons consist of four family members which are the up quark, down quark, electron and its anti–neutrino. There are two other four member families. The TOP family has the highest rest mass energy particle members. The CHARMED family has intermediate rest mass energy particle members. The UP family has the lowest rest mass energy particles. Each family maintains the same relationships between its members.

The UP Family

The UP family consists of an up quark, $u^{2/3}$, a down quark, $d^{-1/3}$, electron, e^-, with its electron anti–neutrino, $\acute{\upsilon}_e$. The quarks can either be red, blue or green. The up quark has a charge of 2/3 e^+ while the down quark has a charge of $-1/3$ e^+. The electron has a rest mass energy of .511 Mev. These particles have the lowest rest mass energy and represent the ground state rest mass energy of the matter families. All UP fermion family members have a spin of $\frac{1}{2}\hbar$ and a helicity of plus or minus $\frac{1}{2}\hbar$.

The CHARMED Family

The CHARMED family consists of a charmed quark, $c^{2/3}$, a strange quark, $s^{-1/3}$, muon, μ^- with its muon anti–neutrino, $\dot{\upsilon}_\mu$. The quarks can either be red, blue or green. The charmed quark has a charge of 2/3 e^+ while the strange quark has a charge of $-1/3$ e^+. The muon has a rest mass energy of 105.66 Mev. These particles have intermediate energy and represent a higher rest mass energy state than the UP family. All CHARMED fermion family members have a spin of $\frac{1}{2}\hbar$ and a helicity of plus or minus $\frac{1}{2}\hbar$.

The TOP Family

The TOP family consists of a top quark, $t^{2/3}$, bottom quark, $b^{-1/3}$, tauon, τ^- with its tauon anti–neutrino, $\acute{\upsilon}_\tau$. The quarks can either be red, blue or green. The top quark has a charge of 2/3 e^+ while the bottom quark has a charge of $-1/3$ e^+. The tauon has a rest mass of 1784.2 Mev. These particles have the highest rest mass energy state and represent a higher energy state than that of the CHARMED family. All TOP fermion family members have a spin of $\frac{1}{2}\hbar$ and a helicity of plus or minus $\frac{1}{2}\hbar$.

Field Energy Particles

Gluon Family

Gluons ($g_1 - g_8$) are responsible for the strong force field between the three colored (red, green and blue) quarks making up protons and neutrons, of which all nuclei are composed. There are eight different gluons. Gluons carry color combinations (r, g, b, –r, –g, –b) and compose the gluon field holding quark trios together in protons and neutrons. Gluons have a spin of $\hbar$. Gluons have zero rest mass and therefore move at c, the velocity of light. Thus, gluons have helicity of either plus or minus $\hbar$.

Photon Family

Photons (γ) are responsible for the electromagnetic forces which act between charges. Photons have no color and no charge. Photons have a spin of $\hbar$. Photons have no rest mass and move at the velocity of light. Thus, photons have helicity of either plus or minus one $\hbar$. The positive helicity photon is the anti–photon of the negative helicity photon. While in flight, photons have mass, energy and momentum.

The Weakon Family

Weakons give rise to the electroweak force field responsible for radioactive decay. Recall that a neutron is composed of two down quarks and one up quark. The decay of an isolated neutron is an example of radioactive beta (electron) decay in which one of the down quarks in a neutron decays into a weakon (the omega minus) which then decays into an up quark, electron and anti–neutrino. The net effect is that a neutron decays into a proton, electron and anti–neutrino. There are three different weakons, the omega minus (Ω^-), omega zero or zeta (Z^0) and the omega plus (Ω^+). These weakons have no color and carry charges of e^-, 0, e^+ respectively. Weakons have a spin of $\hbar$ and each can have helicity of $-\hbar$, 0 or $+\hbar$. Weakons have rest masses of 85 Mp, 260 Mp and 85 Mp respectively. Anti–weakons have opposite charges and helicities as the corresponding weakons.

The Meson Families

Mesons give rise to the forces between baryons (quark trios). Mesons are not elemental but are composed of quark anti–quark pairs (combos taken from any of the three families of quarks) and are mentioned here for completeness. Obviously, there are many families of mesons, and the pi meson family (pions) are responsible for forces between nucleons (either neutrons or protons). Pions will be presented next as an example.

The Pi Meson (Pion) Family

The Pions are responsible for forces between nucleons (either neutrons or protons) and are composed of quark anti–quark pairs. The pi minus (π^-) is composed of a down quark with a charge of $-1/3$ e^+ and an anti–up quark with a charge of $-2/3$ e^+ for a total charge of e^-. The pi zero (π^0) is a mixture of an up quark and an anti-up quark, with a down quark and an anti–down quark. The pi plus (π^+) is composed of an up quark with a charge of $+2/3$ e^+ and an anti–down quark with a charge of $+1/3$ e^+ for a total charge of e^+. These pions have no color and carry charges of e^-, 0, e^+ respectively. Pions have a spin of 0 and each has helicity of 0. The charged pions have rest masses of 139.57 Mev, while the pi zero has a rest mass of 134.96 Mev. The anti–pi minus is the pi plus. The anti–pi plus is the pi minus. The anti pi zero is the pi zero itself.

Graviton Family

Gravitons (G_- and G_+) are responsible for the gravitational force fields which act between masses. Gravitons have no color and no charge.

The G_- graviton has a spin of $2\hbar$ and a zero rest mass. It moves at the velocity of light and thus, its helicity is $-2\hbar$ or $+2\hbar$. It is assumed to have negative mass in flight while being exchanged between any two positive masses or any two negative anti–masses. This is because the gravitational potential energy between two positive masses or two negative masses is negative.

Because of a new scientific theory called "Nature of the First Cause", in which positive matter is gravitationally repelled by negative anti–matter, the G_+ graviton is postulated to exist. It also has a spin of $2\hbar$. It is assumed to have a zero rest mass and moves with the velocity of light and has positive mass in its flight between negative antimatter and positive matter. Thus, the G_+ graviton also has helicity of $-2\hbar$, or $2\hbar$. The G_+ gravitons fill up all space and are responsible for the force of repulsion between negative anti–matter and positive matter. By the "First Cause" theory, it makes up the repulsive gravitational field which is responsible for the accelerated expansion of distant positive matter in the universe (galaxies not in the local group).

The Higgs Family

There are two Higgs bosons (H_L and H_H) called the light Higgs boson, H_L of the unified electroweak theory and the heavy Higgs boson, H_H of the grand unified theory. The heavy Higgs boson, makes up the Higgs field and permeates all space. This field is responsible for assigning masses to all the fundamental particles. The light Higgs boson is responsible for assigning the masses to the weakons. Both Higgs bosons have a spin of zero (0), and thus they both have a helicity of zero. Both Higgs bosons have non–zero rest masses with the light Higgs rest mass at roughly 10^5 Mev and the heavy Higgs rest mass of about 10^{17} Mev.

Complete Set of Particles

All matter particles which have been discovered are combinations of the above elementary matter particles. All the known force fields consists of varying energy and intensity of the above force field particles.

The Hadrons (consisting of quarks) which are matter particles that have been discovered now number over two hundred exceeding the number of known elements.

References

Alexander, Gordon, *Biology*, New York, Barnes & Noble, Inc., 1963

Alexander, Peter, *Atomic Radiation and Life*, Baltimore, Maryland, Penguin Books, 1965

Al-Khalili, Jim, *Quantum, A Guide for the Perplexed*, United Kingdom, Weidenfeld & Nicolson, 2003

Ames, Joseph Sweetman & Murnaghan, Francis D., *Theoretical Mechanics An Introduction to Mathematical Physics*, New York, Dover Publications, Inc., 1957

Anselm, St., *Proslogium*, ed. by Sidney Deane, LaSalle, Illinois, Open Court Publishing Company, 1939

Aquinas, Saint Thomas, *Basic Writings of*, ed. by Anton C. Pegis, 2 Vols, New York, Random House, 1945

Ardrey, Robert, *The Territorial Imperative*, New York, N.Y., Dell Publishing Company, Inc., 1966

Asimov, Isaac, *The Genetic Code*, New York, Signet Books, 1963

Asimov, Isaac, *The Gods Themselves*, New York, Bantam Books, 1972

Atkins, K. R., *Physics*, New York, John Wiley & Sons, Inc., 1965

Ayer, A. J., *On the Analysis of Moral Judgements*, Horizon, 1949

Bennett, Jeffrey & Donahue, Megan & Schneider, Nicholas & Voit, Mark, *The Cosmic Perspective*, New York, Addison Wesley, 2004

Bergmann, Peter Gabriel, *Introduction to the Theory Of Relativity*, New York, Dover Publications, Inc., 1976

Berkeley, George, *Three Dialogues*, New York, Liberal Arts Press, 1954

Blass, Gerhard A., *Theoretical Physics*, New York, Appleton-Century-Crofts, 1962

Bova, Ben, *The Fourth State of Matter*, New York, New American Library, Inc., 1974

Born, Max, *Einstein's Theory of Relativity*, New York, Dover Publications, Inc., 1962

Bradley, F. H., *Appearance and Reality*, Oxford, Clarendon Press, 1930

Breithaupt, Jim, *Cosmology*, Blacklick, OH, McGraw-Hill, 1999

Bronowski, J., *The Ascent of Man*, Boston, Little, Brown and Company, 1973

Burr, Harold Saxton, *The Fields of Life*, New York, Ballantine Books, Inc., 1972

Calder, Nigel, *The Mind of Man*, New York, N.Y., The Viking Press 1973

Darwin, Charles, *The Origin of Species*, New York, Washington Square Press, 1963

Davies, Paul, *The Mind of God: The Scientific Basis for a Rational World*, New York, Touchstone, 1993

Davies, Paul, *The New Physics*, New York, Cambridge University Press, 1996

Dawood, N. J. & Wyatt, Thomas, *The Koran*, London, Penguin Classics, 1990

De Broglie, Louis, *matter and light*, New York, Dover Publications, Inc., 1939

Descartes, Rene, *Meditations on First Philosophy*, in The Philosophical Works of Descartes, New York, Dover, 1955

Dyson, Freeman, *Imagined Worlds*, Cambridge, Mass., Harvard University Press, 1998

Eccles, John C., *Brain and the Unity of Conscious Experience*, Springer-Verlag, 1965

Eddington, Sir A. S., *The Nature of the Physical World*, Folcroft Library Editions, 1935

Eddington, Sir A. S., *Science and the Unseen World*, New York, MacMillan, 1929

Ehrlich, Paul R., *The Population Bomb*, New York, Buccaneer Books, Inc., 1971

Einstein, Albert, *Builders of the Universe*, Los Angeles, CA, U. S. Library Association, Inc., 1932

Einstein, Albert, & Lorentz, H. & A., Minkowski, H., & Weyl, H., *The Principle of Relativity*, New York, Dover Publications, Inc., 1952

Einstein, Albert, *Relativity The Special and General Theory*, New York, Crown Publishers, Inc., 1961

Evans, W. Y. & Wentz, *The Tibetan Book of the Dead*, New York, Oxford University Press, 1978

Fermi, Enrico, *thermodynamics*, New York, Dover Publications Inc., 1956

Feynman, Richard P., *QED The Strange Theory of Light and Matter*, Princeton, New Jersey, Princeton University Press, 1988

Feynman, Richard P., *Six Not So Easy Pieces*, New York, Basic Books, 1997

Frankel, Theodore, *Gravitational Curvature*, San Francisco, W. H. Freeman and Company, 1979

Fromm, Erich, *God and Philosophy*, Yale University Press, 1941

Fuller, R. Buckminster, *Operating Manual for Spaceship Earth*, New York, Aeonian Press, 1976

Gaer, Joseph, *The Wisdom of the Living Religions*, New York, Dodd, Mead & Company, 1956

Gamow, George, *Gravity*, New York, Dover Publications, Inc., 2002

Gandhi, M. K., *The Story of My Experiments With Truth*, Washington, D. C., Public Affairs Press, 1948

Goldstein, Herbert, *Classical Mechanics*, London, Addison-Wesley Publishing Company, Inc., 1950

Goldstein, Kurt, *The Organism*, Boston, Beacon, 1963

Greene, Brian, *The Elegant Universe: Superstrings, Hidden Dimensions, and the Quest for the Ultimate Theory*, New York, W. W. Norton, 1999

Guth, Alan, *The Inflationary Universe,* Reading, Penn., Addison-Wesley, 1997

Haisch, Bernard, *The God Theory*, San Francisco, Ca., Weiser Books, 2006

Halliday, David & Resnick, Robert, *Physics For Students of Science and Engineering*, New York, John Wiley & Sons, Inc., 1962

Hammond, Richard, The Unknown Universe, Franklin Lakes, New Jersey, New Page Books, 2008

Hawking, Stephen & Penrose, Roger, *The Nature of Space and Time*, New Jersey, Princeton University Press, 1996

Hawking, Stephen, *A Brief History of Time: From the Big Bang to Black Holes*, New York, Bantam Books, 1988

Heisenberg, Werner Karl, *The Nature of Elementary Particles*, in Physics Today, Page 39, March 1976

Heisenberg, Werner Karl, *Physics and Beyond: Encounters and Conversations*, New York, Harper and Row, 1971

Herbert, Nick, *Quantum Reality*, New York, Anchor Books, 1987

Hick, John, *The Existence of God*, ed. by Paul Edwards, New York, Macmillan Publishing Company, 1964

Holum, John R., *Fundamentals of General, Organic, and Biological Chemistry*, New York, John Wiley & Sons, 1982

Hubbard, L. Ron, *Dianetics*, Los Angeles, Bridge
Publications, Inc., 1984

Hume, David, *A Treatise of Human Nature*, Selby-
Bigge ed., Oxford University Press, 1951

Hume, David, *Enquiry Concerning Human
Understanding*, Oxford University Press, 1951

Huxley, Aldous, *Brave New World*, New York,
Harper and Row, 1946

Huxley, Aldous, *The Doors of Perception*, New
York, Harper and Row, 1954

Huxley, Thomas Henry, *Evolution and Ethics and
Other Essays*, University Press of the Pacific, 2002

Jackson, John David, *Classical Electrodynamics*,
New York, John Wiley & Sons, Inc., 1963

James, William, *Essays on Faith and Morals*, New
York, Meredian, 1962

James, William, *The Varieties of the Religious Experience,* New York, Modern Library

James, William, *The Will to Believe*, ed. by C. M. Bakewell, Everyman's Library, 1918

Johnson, Raynor C., *Nurslings of Immortality*, New York, N.Y., Harper & Row 1972

Kaku, Michio, *Hyperspace*, New York, Anchor Books, 1995

Kaku, Michio, *Parallel Worlds*, New York, Anchor Books, 2006

Kant, Immanuel, *Prolegemena to any Future Metaphysics*, Chicago, Open Court Publishing Company, 1949

Kant, Immanuel, *Selections*, ed. by T. M. Greene, New York, Chas. Scribners, 1929

Kaplan, Irving, *Nuclear Physics*, Reading, Massachusetts, Addison-Wesley Publishing Company, Inc., 1962

Kawalski, Gary, *Science and the Search for God*, New York, Lantern Books, 2003

Kierkegaard, Soren, *Fear and Trembling*, Garden City, New York, Doubleday, 1955

Kierkegaard, Soren, *Training in Christianity*, London, Oxford University Press, 1941

MacKenzie, Charles, A., *Unified Organic Chemistry*, New York, Harper & Brothers, 1962

Maltz, Dr. Maxwell, *Psycho-Cybernetics*, New York, N.Y., Pocket Books 1969

McMahon, David, *quantum mechanics demystified*, New York, McGraw Hill, 2005

Messiah, Albert, *Quantum Mechanics*, New York, Dover Publications, Inc., 1999

Miller, K. R., *Finding Darwins' God: A Scientist's Search for Common Ground Between God and Evolution*, New York, Harper Collins, 1999

Moody, Dr. Raymond A., Jr., *Life After Life*, New York, Bantam Books 1978

Moyers, Bill, *Healing and the Mind*, New York, Doubleday, 1979

Neumann, John Von, *The Computer and the Brain*, New Haven, Conn., Yale University Press, 1958

Neumann, John Von & Morgenstern, Oskar, *Theory of Games and Economic Behavior*, Princeton University Press, 1943

Nietzche, Fredrich Wilhelm, *The Philosophy of Nietzche*, New York, Modern Library

Park, David, *Introduction to the Quantum Theory*, New York, McGraw-Hill Book Company, 1964

Pauling, Linus, *General Chemistry*, New York, Dover Publications, Inc., 1988

Peebles, P. J. E., *Principles of Physical Cosmology*, Princeton, New Jersey, Princeton University Press, 1993

Penrose, Roger, *The Road To Reality, A Complete Guide to the Laws of the Universe*, New York, Vintage Books, 2004

Planck, Max, *The Universe in the Light of Modern Physics*, G. Allen and Unwin, 1931

Powell, John L. & Crasemann, Bernd, *Quantum Mechanics*, Reading, Massachusetts, Addison-Wesley Publishing Company, Inc., 1961

Rashdall, Hastings, *The Theory of Good and Evil*, London, Oxford University Press, 1907

Ridpath, Ian, *The Illustrated Encyclopedia of the Universe*, New York, Watson-Guptil Publications, 2001

Riggs, Shelton, *The Nature of the First Cause,* Version 11.7, El Paso, Texas, El Paso County Courthouse, 2006

Sahakian, William S., *Ethics: An Introduction to Theories and Problems*, New York, Barnes & Noble Books, 1974

Schopenhauer, Arthur, *The Basis of Morality*, London, Swan Sonnenschien, 1903

Schrodinger, Erwin, *Science and Humanism*, Cambridge University Press, 1951

Schrodinger, Erwin, *What is Life?*, Cambridge University Press, 1944

Schroeder, Gerald L., *The Science of God*, New York, Broadway Books, 1997

Schweitzer, Albert, *The Light Within Us*, New York, Philosophical Library, 1959

Schweitzer, Albert, *The Philosophy of Civilization*, New York, Macmillan, 1951

Sears, Francis W., & Zemansky, Mark W. & Young, Hugh D., *College Physics*, Menlo Park, California, Addison-Wesley Publishing Company, 1986

Segre, Emilio, *Nuclei and Particles*, New York, W. A. Benjamin, Inc., 1964

Setlow, Richard B., & Pollard, Ernest C., *Molecular Biophysics*, London, Addison-Wesley Publishing Company, Inc., 1964

Sholem, G., *Kabbalah*, Marboro Books, 1978

Shortley, George & Williams, Dudley, *Elements of Physics For Students of Science and Engineering*,

Englewood Cliffs, New Jersey, Prentice-Hall, Inc., 1965

Silk, J. *The Big Bang*, New York, W.H. Freeman and Co., 1980

Sohrab, Mizra Ahmad, *The Bible of Mankind*, New York, Universal Publishing Company, 1929

Sorley, W. R., *Moral Values and the Idea of God*, Cambridge University Press, 1918

Spinoza, *God, Man, and Human Welfare*, Chicago, Open Court, 1909

Teilhard de Chardin, P., *The Prayer of the Universe*, New York, Harper Perennial, 1958

Tillich, Paul, *The Concept of God*, in Perspective: A Princeton Journal of Christian Opinion, II No. 3, January 1950

Van Heuvelen, Alan, *Physics, A General Introduction*, Boston, Little, Brown and Company, 1982

Walsch, N. D., *Conversations with God: An Uncommon Dialog*, New York, G.P. Putnam's Sons, 1996

Watson, James D., *The Double Helix*, New York, The New American Library, Inc., 1968

Weinberg, Steve*, Dreams of a Final Theory: The Search for the Fundamental Laws of Nature*, New York, Pantheon Books, 1992

Weld, LeRoy D., *A Textbook of Heat*, New York, The Macmillan Company, 1948

Winchester, A. M., *Heredity An Introduction To Genetics*, New York, Barnes & Noble, Inc., 1966

Weyl, Hermann, *Symmetry*, Princeton University Press, 1952

Whitehead, Alfred North, *Religion in the Making*, New York, Macmillan, 1926

Wooldridge, Dean E., *The Machinery of the Brain*, New York, McGraw-Hill Book Company, Inc., 1963

Zaffuto, Dr. Anthony A. & Mary Q., *Alpha-genics*, New York, N.Y., Warner Paperback Library 1975

Zondervan Publishing House, *The Holy Bible New International Version*, Grand Rapids, Michigan, 1988

INDEX

A

B

C

F

L

M

O

P

R

S

T